STANDARD GRADE | CREDIT

2008

[BLANK PAGE]

OFFICIAL SQA PAST PAPERS WITH ANSWERS

STANDARD GRADE | CREDIT

CHEMISTRY
2008-2012

2008 EXAM — page 3
2009 EXAM — page 33
2010 EXAM — page 63
2011 EXAM — page 89
2012 EXAM — page 121
ANSWER SECTION — page 151

BrightRED
PUBLISHING

First exam published in 2008.
Published by Bright Red Publishing Ltd, 6 Stafford Street, Edinburgh EH3 7AU
tel: 0131 220 5804 fax: 0131 220 6710 info@brightredpublishing.co.uk www.brightredpublishing.co.uk

ISBN 978-1-84948-239-4

A CIP Catalogue record for this book is available from the British Library.

Bright Red Publishing is grateful to the copyright holders, as credited on the final page of the Question Section, for permission to use their material. Every effort has been made to trace the copyright holders and to obtain their permission for the use of copyright material. Bright Red Publishing will be happy to receive information allowing us to rectify any error or omission in future editions.

FOR OFFICIAL USE

C

KU   PS

Total Marks

0500/402

NATIONAL
QUALIFICATIONS
2008

THURSDAY, 1 MAY
10.50 AM – 12.20 PM

CHEMISTRY
STANDARD GRADE
Credit Level

Fill in these boxes and read what is printed below.

Full name of centre

Town

Forename(s)

Surname

Date of birth

Day   Month   Year   Scottish candidate number   Number of seat

1 All questions should be attempted.

2 Necessary data will be found in the Data Booklet provided for Chemistry at Standard Grade and Intermediate 2.

3 The questions may be answered in any order but all answers are to be written in this answer book, and must be written clearly and legibly in ink.

4 Rough work, if any should be necessary, as well as the fair copy, is to be written in this book.

   Rough work should be scored through when the fair copy has been written.

5 Additional space for answers and rough work will be found at the end of the book.

6 The size of the space provided for an answer should not be taken as an indication of how much to write. It is not necessary to use all the space.

7 Before leaving the examination room you must give this book to the invigilator. If you do not, you may lose all the marks for this paper.

PART 1

In Questions 1 to 9 of this part of the paper, an answer is given by circling the appropriate letter (or letters) in the answer grid provided.

In some questions, two letters are required for full marks.

If more than the correct number of answers is given, marks will be deducted.

A total of 20 marks is available in this part of the paper.

SAMPLE QUESTION

A		B		C	
	CH_4		H_2		CO_2
D		E		F	
	CO		C_2H_5OH		C

(a) Identify the hydrocarbon.

Ⓐ	B	C
D	E	F

The one correct answer to part (a) is A. This should be circled.

(b) Identify the **two** elements.

A	Ⓑ	C
D	E	Ⓕ

As indicated in this question, there are **two** correct answers to part (b). These are B and F. Both answers are circled.

If, after you have recorded your answer, you decide that you have made an error and wish to make a change, you should cancel the original answer and circle the answer you now consider to be correct. Thus, in part (a), if you want to change an answer A to an answer D, your answer sheet would look like this:

(A̶)	B	C
Ⓓ	E	F

If you want to change back to an answer which has already been scored out, you should enter a tick (✓) in the box of the answer of your choice, thus:

✓(A̶)	B	C
(D̶)	E	F

Marks | KU | PS

1. The formulae of some gases are shown in the grid.

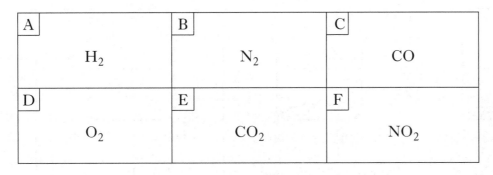

A	B	C
H_2	N_2	CO
D	E	F
O_2	CO_2	NO_2

(a) Identify the toxic gas produced during the burning of plastics.

A	B	C
D	E	F

1

(b) Identify the gas which makes up approximately 80% of air.

A	B	C
D	E	F

1

(c) Identify the gas used up during respiration.

A	B	C
D	E	F

1

(3)

[Turn over

Marks KU PS

2. A student carried out several experiments with metals and acids.

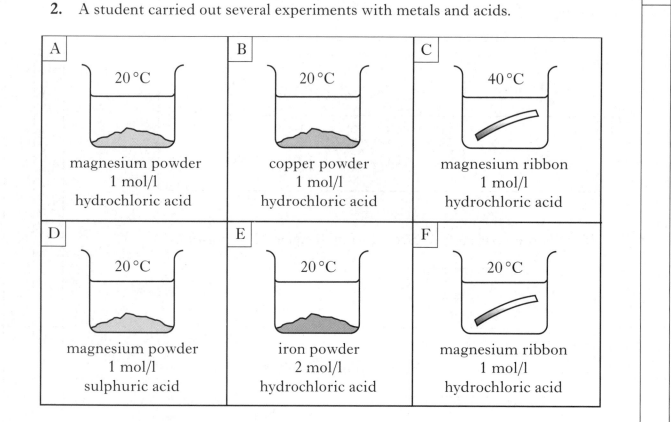

A 20 °C magnesium powder 1 mol/l hydrochloric acid	B 20 °C copper powder 1 mol/l hydrochloric acid	C 40 °C magnesium ribbon 1 mol/l hydrochloric acid
D 20 °C magnesium powder 1 mol/l sulphuric acid	E 20 °C iron powder 2 mol/l hydrochloric acid	F 20 °C magnesium ribbon 1 mol/l hydrochloric acid

(a) Identify the **two** experiments which could be compared to show the effect of particle size on reaction rate.

A	B	C
D	E	F

1

(b) Identify the experiment in which **no** reaction would take place.

A	B	C
D	E	F

1

(2)

Marks KU PS

3. The grid shows the structural formulae of some hydrocarbons.

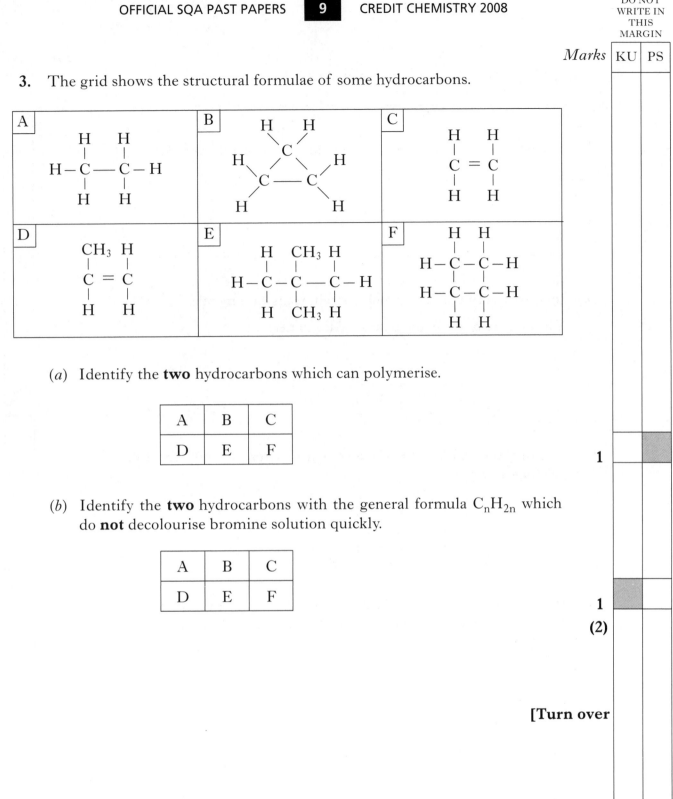

(a) Identify the **two** hydrocarbons which can polymerise.

A	B	C
D	E	F

1

(b) Identify the **two** hydrocarbons with the general formula C_nH_{2n} which do **not** decolourise bromine solution quickly.

A	B	C
D	E	F

1

(2)

[Turn over

Marks | KU | PS

4. The grid shows the names of some oxides.

A	B	C
silicon dioxide	carbon dioxide	sodium oxide
D	E	F
iron oxide	sulphur dioxide	copper oxide

(a) Identify the **two** oxides which contain transition metals.

You may wish to use the data booklet to help you.

A	B	C
D	E	F

1

(b) Identify the oxide which reacts with water in the atmosphere to produce acid rain.

A	B	C
D	E	F

1

(c) Identify the oxide which, when added to water, produces a solution with a greater concentration of hydroxide ions (OH^-) than hydrogen ions (H^+).

A	B	C
D	E	F

1

(3)

Marks | KU | PS

5. There are different types of chemical reactions.

A	redox
B	precipitation
C	combustion
D	neutralisation
E	displacement

(a) Identify the type of chemical reaction taking place when dilute hydrochloric acid reacts with a carbonate.

A
B
C
D
E

1

(b) Identify the **two** types of chemical reaction represented by the following equation.

$$2Zn(s) + O_2(g) \longrightarrow 2ZnO(s)$$

A
B
C
D
E

2

(3)

[Turn over

Marks | KU | PS

6. Lemonade can be made by dissolving sugar, lemon and carbon dioxide in water.

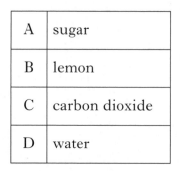

A	sugar
B	lemon
C	carbon dioxide
D	water

Identify the solvent used to make lemonade.

A
B
C
D

(1)

DO NOT WRITE IN THIS MARGIN

Marks | KU | PS

7. The grid contains the names of some carbohydrates.

A	fructose
B	glucose
C	maltose
D	sucrose
E	starch

(*a*) Galactose is a monosaccharide found in dairy products.

Identify the **two** isomers of galactose.

A
B
C
D
E

1

(*b*) Identify the carbohydrate which is a condensation polymer.

A
B
C
D
E

1

(2)

[Turn over

8. A student made some statements about acids.

A	Acid rain will have no effect on iron structures.
B	A base is a substance which can neutralise an alkali.
C	Treatment of acid indigestion is an example of neutralisation.
D	In a neutralisation reaction the pH of the acid will fall towards 7.
E	When dilute nitric acid reacts with potassium hydroxide solution, the salt potassium nitrate is produced.

Identify the **two** correct statements.

A
B
C
D
E

(2)

Marks | KU | PS

9. Coffee manufacturers have produced a self-heating can of coffee.

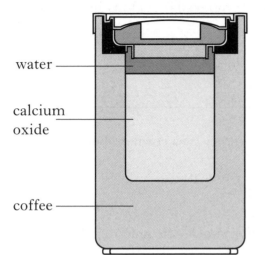

water —
calcium
oxide —
coffee —

In the centre of the can calcium oxide reacts with water, releasing heat energy.

The equation for the reaction is:

$$CaO(s) + H_2O(\ell) \longrightarrow Ca(OH)_2(aq)$$

A	Calcium oxide is insoluble.
B	The reaction is exothermic.
C	The reaction produces an acidic solution.
D	The temperature of the coffee goes down.
E	0·1 moles of calcium oxide reacts with water producing 0·1 moles of calcium hydroxide.

Identify the **two** correct statements.

A
B
C
D
E

(2)

Marks | KU | PS

PART 2

A total of 40 marks is available in this part of the paper.

10. Hydrogen reacts with other elements to form molecules such as hydrogen fluoride and hydrogen chloride.

    (a) Name the family to which fluorine and chlorine belong.

    _____   **1**

    (b) The atoms in these molecules are held together by a covalent bond.

    Circle the correct words to complete the sentence.

    A covalent bond forms when two $\left\{\begin{array}{l}\text{positive}\\\text{negative}\\\text{neutral}\end{array}\right\}$ nuclei are held together by

    their common attraction for a shared pair of $\left\{\begin{array}{l}\text{protons}\\\text{neutrons}\\\text{electrons}\end{array}\right\}$.   **1**

    (c) The table gives information about some molecules.

Molecule H–X	Size of X/pm	Energy to break bond kJ/mol
H–F	71	569
H–Cl	99	428
H–Br	114	362
H–I	133	295

    Describe how the size of element **X** affects the energy needed to break the bond in the molecule.

    _____

    _____   **1**

    (3)

Marks | KU | PS

11. Crude oil can be transported to a refinery through a steel pipeline.

(a) If the pipeline is not protected the iron will rust.

Name the **ion** formed from water and oxygen, when they accept electrons during rusting.

_____

1

(b) Some parts of the pipeline are under the sea.

What effect would seawater have on the rate of rusting?

_____

_____

1

(c) Magnesium can be attached to the steel pipeline to prevent rusting.

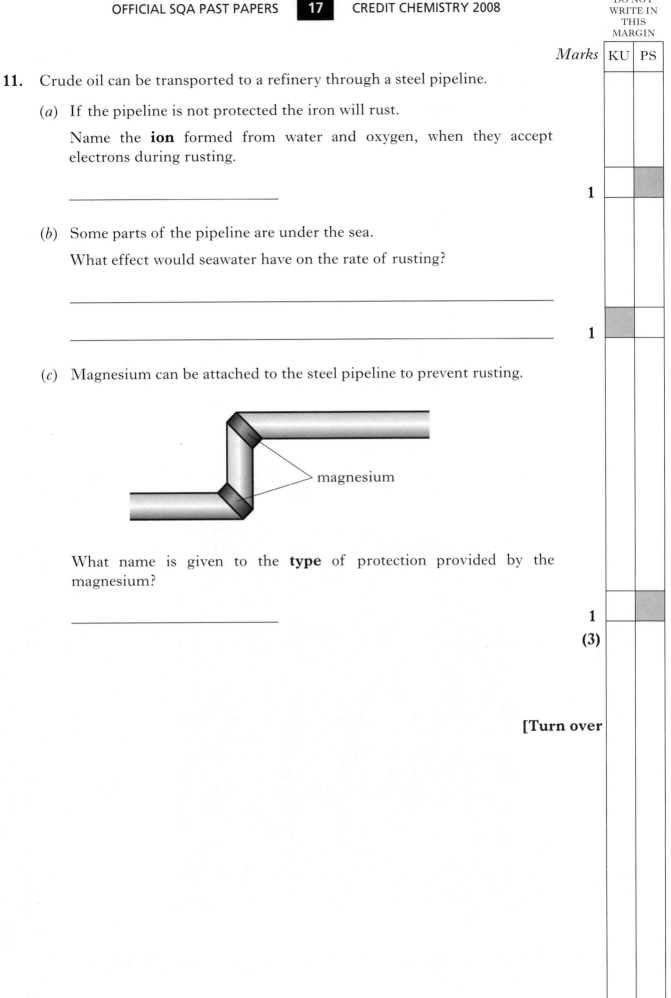

magnesium

What name is given to the **type** of protection provided by the magnesium?

_____

1

(3)

[Turn over

Marks | KU | PS

12. Airbags in cars are designed to prevent injuries in car crashes.

They contain sodium azide (NaN_3) which produces nitrogen gas on impact.

The nitrogen inflates the airbag very quickly.

(*a*) The table gives information on the volume of nitrogen gas produced.

Time/microseconds	Volume of nitrogen gas produced/litres
0	0
5	46
10	64
15	74
20	82
25	88
30	88

(i) Draw a line graph of the results.

Use appropriate scales to fill most of the graph paper.

(Additional graph paper, if required, will be found on page 28.)

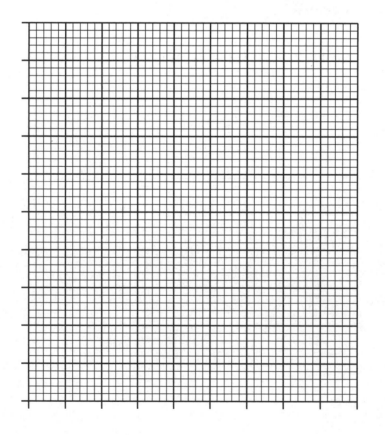

2

(ii) Using your graph, predict the time taken to produce 70 litres of nitrogen gas.

_____ microseconds

1

Marks | KU | PS

12. **(continued)**

(b) The equation for the production of nitrogen gas is:

$$NaN_3(s) \longrightarrow N_2(g) + Na(s).$$

Balance the equation above.

1

(c) Nitrogen is a non-toxic gas.

Suggest another property of nitrogen which makes it a suitable gas for use in airbags.

_____

_____

1

(5)

[Turn over

Marks | KU | PS

13. Copper chloride solution can be broken up into its elements by passing electricity through it.

(a) Carbon is unreactive and insoluble in water.

Give another reason why it is suitable for use as an electrode.

_____ 1

(b) Chlorine gas is released at the positive electrode.

Write an ion-electron equation for the formation of chlorine.
You may wish to use the data booklet to help you.

_____ 1

(c) Why do ionic compounds, like copper chloride, conduct electricity when in solution?

_____

_____ 1

(3)

Marks | KU | PS

14. A fizzy drink "Fizz Alive" contains a carbohydrate.

(a) Name all the elements found in a carbohydrate.

_____

1

(b) A student carried out an investigation to find out which carbohydrate was present in "Fizz Alive".

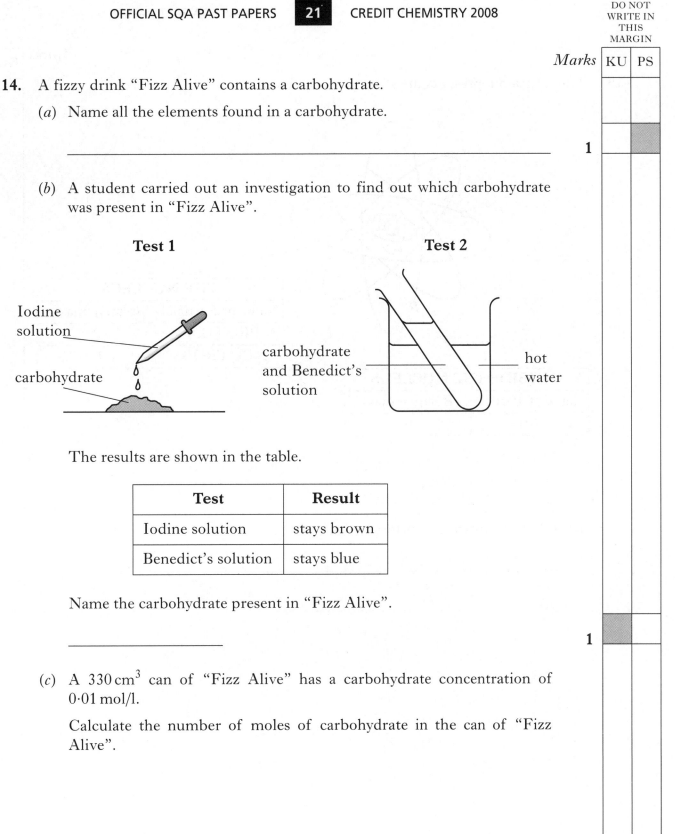

Test 1

Iodine solution

carbohydrate

Test 2

carbohydrate and Benedict's solution

hot water

The results are shown in the table.

Test	Result
Iodine solution	stays brown
Benedict's solution	stays blue

Name the carbohydrate present in "Fizz Alive".

_____

1

(c) A 330 cm^3 can of "Fizz Alive" has a carbohydrate concentration of 0·01 mol/l.

Calculate the number of moles of carbohydrate in the can of "Fizz Alive".

_____ mol

1

(3)

[Turn over

Marks KU PS

15. The diagram represents the structure of an atom.

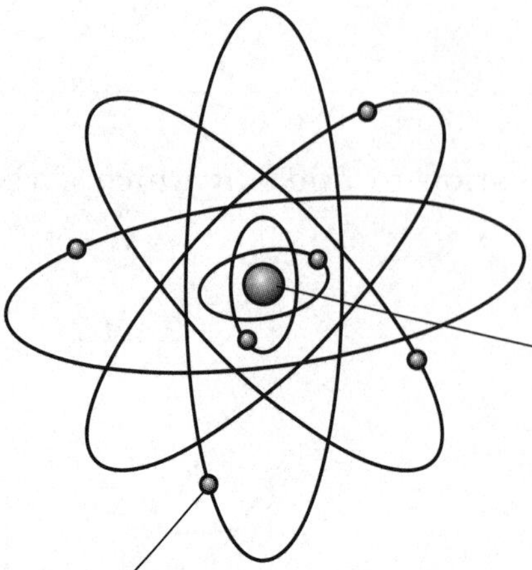

THE NUCLEUS	
Name of Particle	Relative mass
PROTON	(i)
NEUTRON	1

OUTSIDE THE NUCLEUS	
Name of Particle	Relative mass
(ii)	0

(*a*) Fill in the missing information for:

(i) ________________________________

(ii) ________________________________ **1**

15. (continued)

(b) The element uranium has unstable atoms.

These atoms give out radiation and a new element is formed.

$$^{238}_{92}U \longrightarrow \ ^{234}_{90}Th + \ ^{4}_{2}\alpha$$
radiation

(i) Complete the table to show the number of each type of particle in $^{234}_{90}Th$.

Particle	Number
proton	
neutron	

(ii) Radon is another element which gives out radiation.

$$^{222}_{86}Rn \longrightarrow X + \ ^{4}_{2}\alpha$$
radiation

State the **atomic number** of element **X**.

_____

1

1

(3)

[Turn over

Marks | KU | PS

16. Anglesite is an ore containing lead(II) sulphate, $PbSO_4$.

(a) Calculate the percentage by mass of lead in anglesite.

_____ %   **2**

(b) Most metals are found combined in the Earth's crust and have to be extracted from their ores.

Place the following metals in the correct space in the table.

lead   aluminium

You may wish to use the data booklet to help you.

Metal	Method of extraction
	electrolysis of molten compound
	using heat and carbon

1

(c) Metal **X** can be extracted from its ore by heat alone.

What does this indicate about the reactivity of **X** compared to both lead and aluminium?

_____

_____   **1**

(d) When a metal is extracted from its ore, metal ions are changed to metal atoms.

Name this **type** of chemical reaction.

_____   **1**

(5)

Marks KU PS

17. A student added strips of magnesium to solutions of other metals.

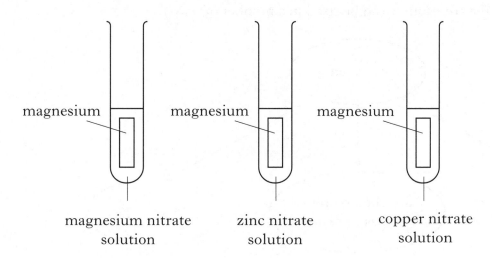

magnesium	magnesium	magnesium
magnesium nitrate solution	zinc nitrate solution	copper nitrate solution

The results are shown in the table.

Solution / Metal	magnesium nitrate	zinc nitrate	copper nitrate
magnesium	(i)	(ii)	reaction occurred

(*a*) In the table, fill in the missing information at (i) and (ii) to show whether or not a chemical reaction has occurred.

You may wish to use the data booklet to help you.   1

(*b*) The equation for the reaction between magnesium and copper nitrate is:

$$Mg(s) + Cu^{2+}(aq) + 2NO_3^-(aq) \longrightarrow Mg^{2+}(aq) + 2NO_3^-(aq) + Cu(s).$$

(i) Circle the spectator ion in the above equation.   1

(ii) What technique could be used to remove copper from the mixture?

_____

_____   1

(3)

[Turn over

Marks | KU | PS

18. Nitrogen is essential for healthy plant growth.

Nitrogen from the atmosphere can be fixed in a number of ways.

(a) **X** is a natural process which takes place in the atmosphere, producing nitrogen dioxide gas.

What provides the energy for this process?

_____ 1

(b) What is present in the root nodules of some plants which convert nitrogen from the atmosphere into nitrogen compounds?

_____

_____ 1

(c) The Haber Process is the industrial method of converting nitrogen into a nitrogen compound.

Name the nitrogen compound produced.

_____ 1

Marks KU PS

18. (continued)

(*d*) The nitrogen compound produced in the Haber Process dissolves in water.

The graph shows the solubility of the nitrogen compound at different temperatures.

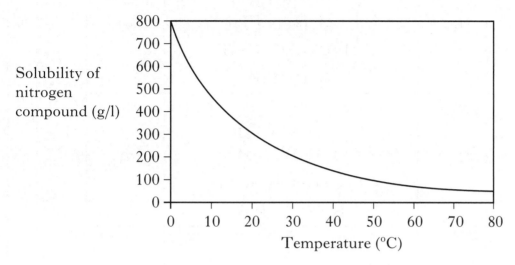

Solubility of nitrogen compound (g/l)

Write a general statement describing the effect of temperature on the solubility of the nitrogen compound.

_____

_____

1
(4)

[Turn over

DO NOT WRITE IN THIS MARGIN

Marks KU PS

19. The octane number indicates how efficiently a fuel burns.

Alkane	Molecular Formula	Full Structural Formula	Octane Number
2-methylbutane	C_5H_{12}	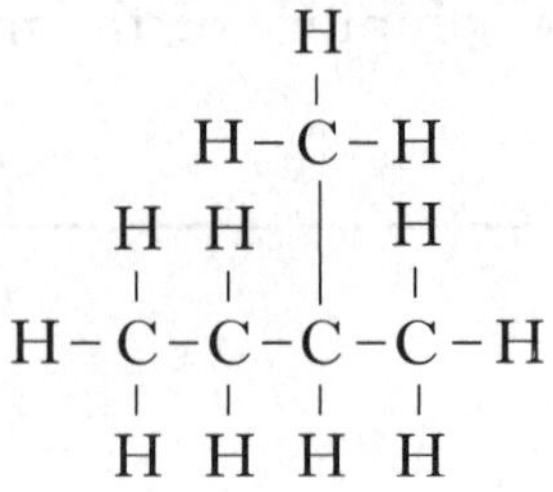	93
2-methylpentane	C_6H_{14}	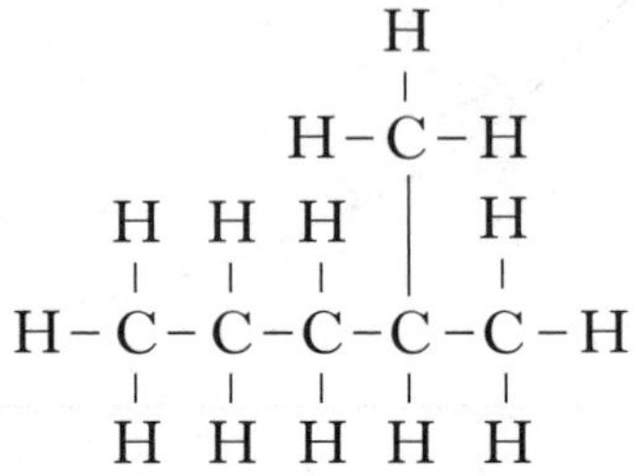	71
2-methylhexane	C_7H_{16}		47
2-methylheptane	C_8H_{18}	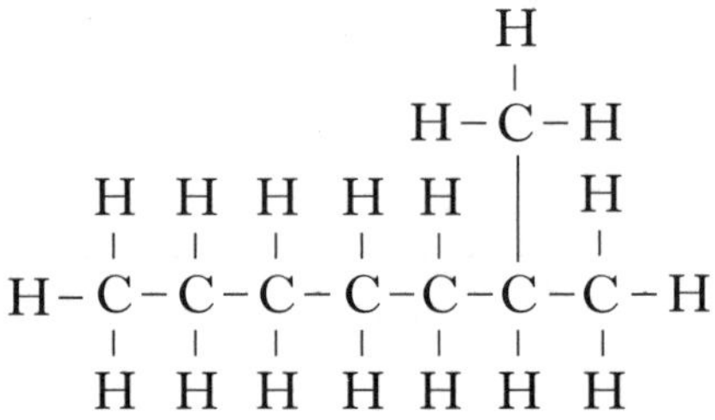	
2-methyloctane	C_9H_{20}	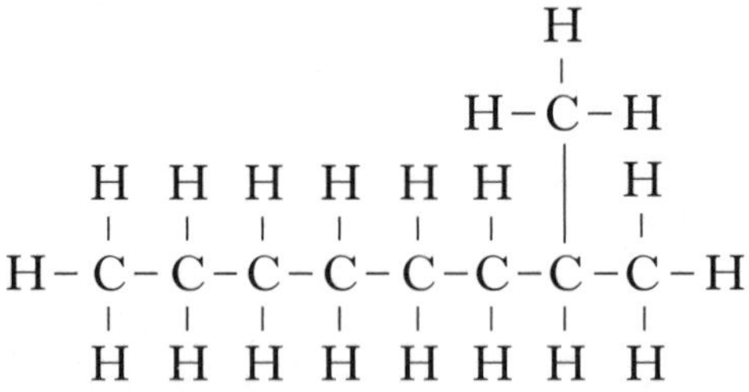	2

(*a*) Draw the **full** structural formula for 2-methylhexane. **1**

Marks

19. (continued)

(b) 2-methylpentane and hexane have the same molecular formula (C_6H_{14}), but different structural formulae.

What term is used to describe this pair of alkanes?

_____   1

(c) Using information in the table, predict the octane number for 2-methylheptane.

_____   1

(3)

[Turn over

DO NOT WRITE IN THIS MARGIN

Marks KU PS

20. Molten iron is used to join steel railway lines together.

Molten iron is produced when aluminium reacts with iron oxide.

The equation for the reaction is:

$$2Al + Fe_2O_3 \longrightarrow 2Fe + Al_2O_3$$

(*a*) Calculate the mass of iron produced from 40 grams of iron oxide.

____________ g **2**

(*b*) The formula for iron oxide is Fe_2O_3.

What is the charge on this iron ion?

____________ **1**

Marks KU PS

20. (continued)

(*c*) Iron can also be produced from iron ore, Fe_2O_3, in a blast furnace.

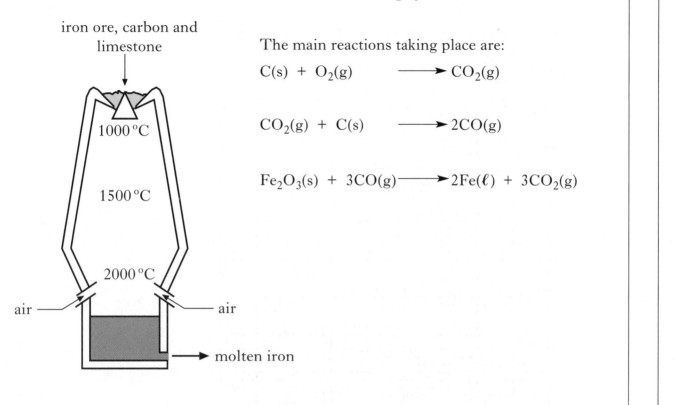

iron ore, carbon and limestone

1000 °C

1500 °C

2000 °C

air — air

→ molten iron

The main reactions taking place are:

$$C(s) + O_2(g) \longrightarrow CO_2(g)$$

$$CO_2(g) + C(s) \longrightarrow 2CO(g)$$

$$Fe_2O_3(s) + 3CO(g) \longrightarrow 2Fe(\ell) + 3CO_2(g)$$

(i) When air is blown into the furnace the temperature rises.

Suggest another reason why **air** is blown into the furnace.

_____

_____   1

(ii) Explain why the temperature at the bottom of the blast furnace should **not** drop below 1535 °C.

You may wish to use the data booklet to help you.

_____

_____   1

(5)

[END OF QUESTION PAPER]

ADDITIONAL SPACE FOR ANSWERS

ADDITIONAL GRAPH PAPER FOR QUESTION 12(*a*)(i)

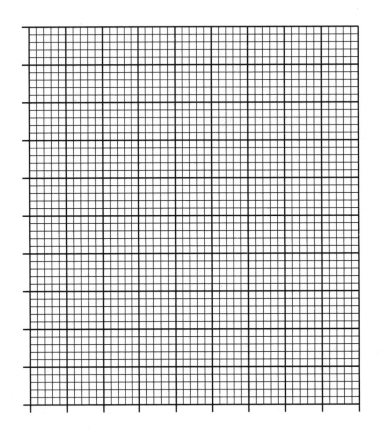

STANDARD GRADE | CREDIT

2009

[BLANK PAGE]

FOR OFFICIAL USE

C

	KU	PS
Total Marks		

0500/402

NATIONAL
QUALIFICATIONS
2009

MONDAY, 11 MAY
10.50 AM – 12.20 PM

CHEMISTRY
STANDARD GRADE
Credit Level

Fill in these boxes and read what is printed below.

Full name of centre

Town

Forename(s)

Surname

Date of birth

Day Month Year   Scottish candidate number   Number of seat

1 All questions should be attempted.

2 Necessary data will be found in the Data Booklet provided for Chemistry at Standard Grade and Intermediate 2.

3 The questions may be answered in any order but all answers are to be written in this answer book, and must be written clearly and legibly in ink.

4 Rough work, if any should be necessary, as well as the fair copy, is to be written in this book.

   Rough work should be scored through when the fair copy has been written.

5 Additional space for answers and rough work will be found at the end of the book.

6 The size of the space provided for an answer should not be taken as an indication of how much to write.   It is not necessary to use all the space.

7 Before leaving the examination room you must give this book to the invigilator. If you do not, you may lose all the marks for this paper.

PART 1

In Questions 1 to 8 of this part of the paper, an answer is given by circling the appropriate letter (or letters) in the answer grid provided.

In some questions, two letters are required for full marks.

If more than the correct number of answers is given, marks will be deducted.

A total of 20 marks is available in this part of the paper.

SAMPLE QUESTION

A CH_4	B H_2	C CO_2
D CO	E C_2H_5OH	F C

(a) Identify the hydrocarbon.

Ⓐ	B	C
D	E	F

The one correct answer to part (a) is A. This should be circled.

(b) Identify the **two** elements.

A	Ⓑ	C
D	E	Ⓕ

As indicated in this question, there are **two** correct answers to part (b). These are B and F. Both answers are circled.

If, after you have recorded your answer, you decide that you have made an error and wish to make a change, you should cancel the original answer and circle the answer you now consider to be correct. Thus, in part (a), if you want to change an answer A to an answer D, your answer sheet would look like this:

(A̸)	B	C
Ⓓ	E	F

If you want to change back to an answer which has already been scored out, you should enter a tick (✓) in the box of the answer of your choice, thus:

✓(A̸)	B	C
(D̸)	E	F

DO NOT WRITE IN THIS MARGIN

Marks | KU | PS

1. Many solutions are used for chemical tests.

A	B	C
Benedict's reagent	lime water	bromine solution
D	E	F
pH indicator	iodine solution	ferroxyl indicator

(a) Identify the solution which could be used to test for maltose.

A	B	C
D	E	F

1

(b) Identify the solution which is used to test for $Fe^{2+}(aq)$.

A	B	C
D	E	F

1

(2)

[Turn over

Marks | KU | PS

2. Many chemical compounds contain ions.

A	B	C
strontium chloride	lithium oxide	calcium oxide
D	E	F
barium fluoride	sodium fluoride	potassium chloride

(a) Identify the compound which produces a green flame colour.

You may wish to use the data booklet to help you.

A	B	C
D	E	F

1

(b) Identify the compound in which **both** ions have the same electron arrangement as argon.

A	B	C
D	E	F

1
(2)

Marks | KU | PS

3. The table contains information about some substances.

Substance	Melting point/°C	Boiling point/°C	Conducts as	
			a solid	a liquid
A	639	3228	yes	yes
B	2967	3273	no	no
C	159	211	no	no
D	1402	2497	no	yes
E	27	677	yes	yes

(a) Identify the substance which exists as a covalent network.

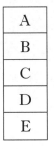

1

(b) Identify the substance which could be calcium fluoride.

A
B
C
D
E

1

(2)

[Turn over

Marks

4. The grid shows the names of some ionic compounds.

A	B	C
aluminium bromide	sodium chloride	potassium hydroxide
D	E	F
sodium sulphate	potassium bromide	calcium chloride

(a) Identify the base.

A	B	C
D	E	F

1

(b) Identify the **two** compounds whose solutions would form a precipitate when mixed.

You may wish to use the data booklet to help you.

A	B	C
D	E	F

1

(c) Identify the compound with a formula of the type XY_2, where **X** is a metal.

A	B	C
D	E	F

1

(3)

Marks | KU | PS

5. The names of some hydrocarbons are shown in the grid.

A ethane	B pentene	C cyclohexane
D pentane	E cyclopentane	F propene

(a) Identify the **two** isomers.

A	B	C
D	E	F

1

(b) Identify the hydrocarbon with the highest boiling point.

You may wish to use the data booklet to help you.

A	B	C
D	E	F

1

(c) Identify the **two** hydrocarbons which can take part in an addition reaction with hydrogen.

A	B	C
D	E	F

1

(3)

[Turn over

6. Reactions can be represented using chemical equations.

A	$Fe^{2+}(aq) + 2e^- \rightarrow Fe(s)$
B	$Fe^{2+}(aq) \rightarrow Fe^{3+}(aq) + e^-$
C	$2H_2(g) + O_2(g) \rightarrow 2H_2O(g)$
D	$2H_2O(\ell) + O_2(g) + 4e^- \rightarrow 4OH^-(aq)$
E	$SO_2(g) + H_2O(\ell) \rightarrow 2H^+(aq) + SO_3^{2-}(aq)$

(a) Identify the equation which shows the formation of acid rain.

A
B
C
D
E

1

(b) Identify the equation which represents a combustion reaction.

A
B
C
D
E

1

(c) Identify the **two** equations which are involved in the corrosion of iron.

A
B
C
D
E

2

(4)

DO NOT WRITE IN THIS MARGIN

Marks KU PS

7. The grid contains information about the particles found in atoms.

A	B	C
relative mass = 1	charge = zero	relative mass almost zero
D	**E**	**F**
charge = 1–	found outside the nucleus	charge = 1+

Identify the **two** terms which can be applied to protons.

A	B	C
D	E	F

(2)

[Turn over

Marks KU | PS

8. The fractional distillation of crude oil was demonstrated to a class.

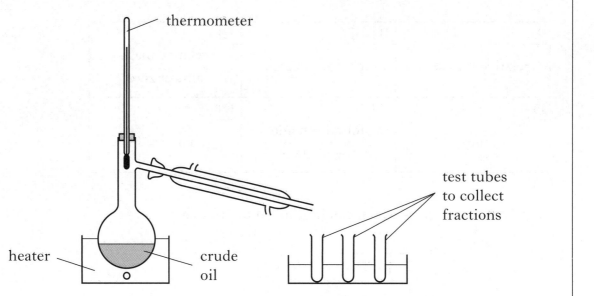

Six fractions were numbered in the order they were collected.

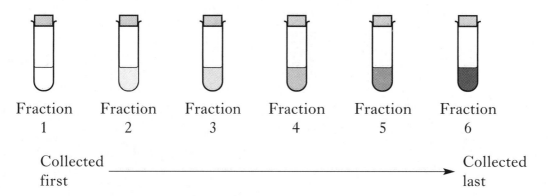

| Fraction 1 | Fraction 2 | Fraction 3 | Fraction 4 | Fraction 5 | Fraction 6 |

Collected first ⟶ Collected last

Identify the **two** correct statements.

A	Fraction 6 evaporates most easily.
B	Fraction 5 is less viscous than fraction 4.
C	Fraction 2 is more flammable than fraction 3.
D	Fraction 1 has a lower boiling range than fraction 2.
E	The molecules in fraction 3 are larger than those in fraction 4.

A
B
C
D
E

(2)

[Turn over for Part 2 on *Page twelve*

PART 2

A total of 40 marks is available in this part of the paper.

9. There are three different types of neon atom.

Type of atom	Number of protons	Number of neutrons
$^{20}_{10}Ne$		
$^{21}_{10}Ne$		
$^{22}_{10}Ne$		

(a) Complete the table to show the number of protons and neutrons in each type of neon atom.

1

(b) What term is used to describe these different types of neon atom?

_____

1

(c) A natural sample of neon has an average atomic mass of 20·2.

What is the mass number of the most common type of atom in the sample of neon?

_____

1

(3)

Marks KU PS

10. Aluminium metal can be produced by passing electricity through molten aluminium oxide.

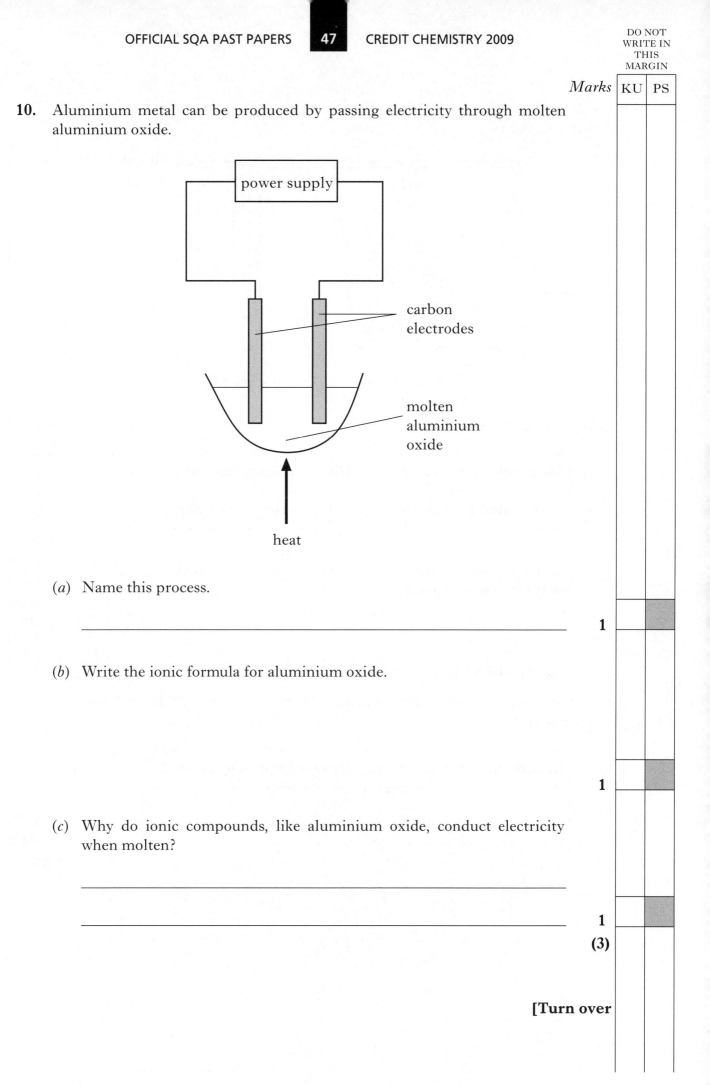

(*a*) Name this process.

_____ **1**

(*b*) Write the ionic formula for aluminium oxide.

1

(*c*) Why do ionic compounds, like aluminium oxide, conduct electricity when molten?

_____

_____ **1**

(3)

[Turn over

Marks | KU | PS

11. A student burned gas **X** and the products were passed through the apparatus shown.

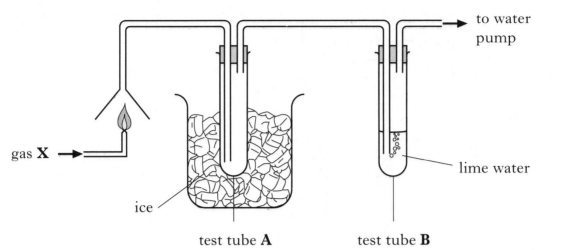

gas **X** →

ice

test tube **A**

to water pump

lime water

test tube **B**

(*a*) The results are shown in the table.

Observation in test tube A	Observation in test tube B
colourless liquid forms	lime water turns milky

Using the information in the table, name two **elements** which **must** be present in gas **X**.

_____   1

(*b*) The experiment was repeated using hydrogen gas.

Complete the table showing the results which would have been obtained.

Observation in test tube A	Observation in test tube B

1

(2)

Marks | KU | PS

12. Hydrogen can form bonds with other elements.

The diagram shows the arrangement of outer electrons in a molecule of hydrogen chloride.

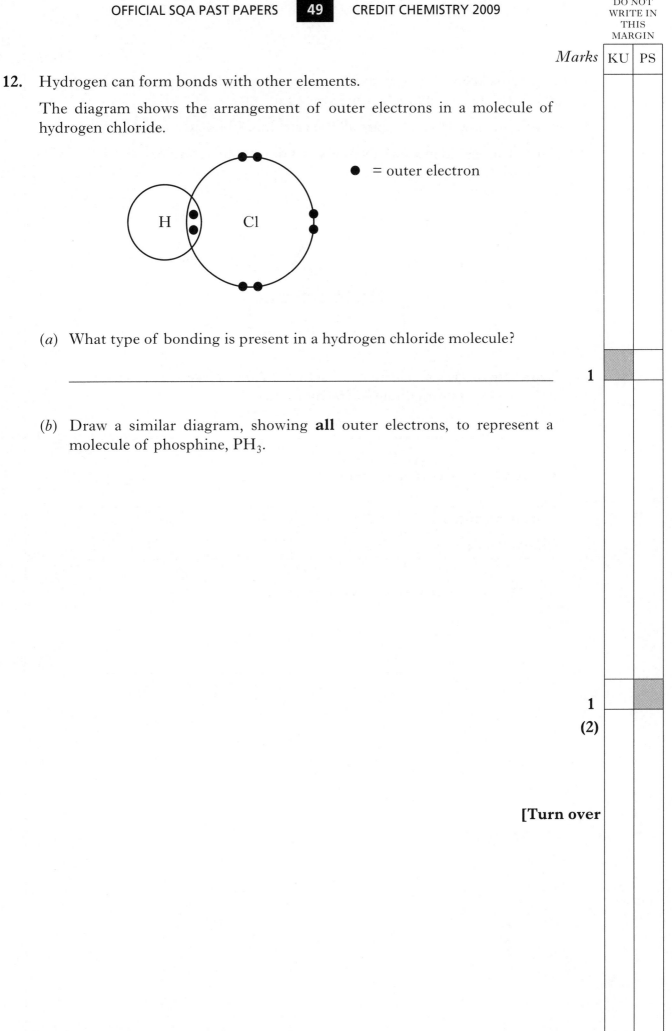

● = outer electron

(a) What type of bonding is present in a hydrogen chloride molecule?

_____

1

(b) Draw a similar diagram, showing **all** outer electrons, to represent a molecule of phosphine, PH_3.

1

(2)

[Turn over

Marks KU PS

13. The apparatus below was used to investigate the reaction between lumps of calcium carbonate and dilute hydrochloric acid.

Excess acid was used to make sure all the calcium carbonate reacted.

A balance was used to measure the mass lost during the reaction.

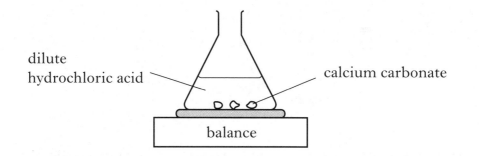

dilute hydrochloric acid

calcium carbonate

balance

(a) Name the type of chemical reaction taking place when calcium carbonate reacts with dilute hydrochloric acid.

_____   1

(b) The results are shown in the table.

Time/minutes	0	0·5	1·0	2·0	3·0	4·0	5·0
Mass lost/g	0	0·36	0·52	0·70	0·80	0·86	0·86

(i) Why is mass lost during the reaction?

_____

_____   1

Marks | KU | PS

13. **(b)** **(continued)**

(ii) Draw a line graph of the results.

Use appropriate scales to fill most of the graph paper.

(Additional graph paper, if required, will be found on page 26.)

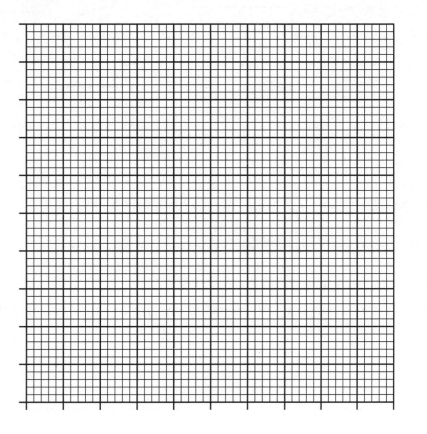

2

(c) The experiment was repeated using the same volume and concentration of acid. The same mass of calcium carbonate was used but **powder** instead of lumps.

Suggest how much mass would have been lost after three minutes.

 g   1

(5)

[Turn over

Marks

14. (*a*) The flow diagram shows how ammonia is converted to nitric acid.

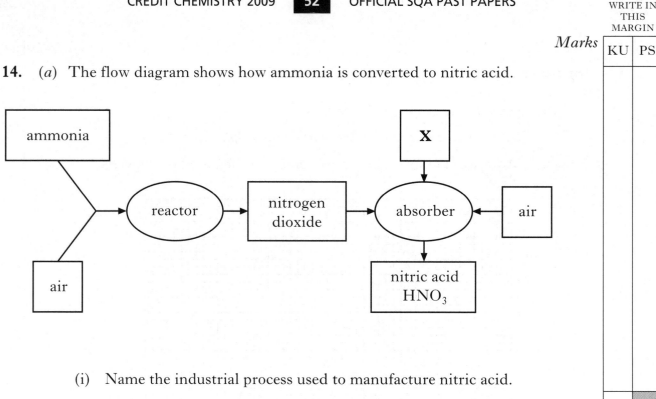

(i) Name the industrial process used to manufacture nitric acid.

_____   1

(ii) The reactor contains a platinum catalyst.

Why is it **not** necessary to continue heating the catalyst once the reaction has started?

_____

_____   1

(iii) Name substance **X**.

_____   1

(*b*) Ammonia and nitric acid react together to form ammonium nitrate, NH_4NO_3.

Calculate the percentage by mass of nitrogen in ammonium nitrate.

Show your working clearly.

_____%   **2**

(5)

15. A student carried out some experiments with four metals and their oxides. The results are shown in the table.

Metal	Reaction with cold water	Reaction with dilute acid	Effect of heat on metal oxide
W	no reaction	no reaction	no reaction
X	no reaction	gas produced	no reaction
Y	gas produced	gas produced	no reaction
Z	no reaction	no reaction	metal produced

(a) Place the four metals in order of reactivity (**most reactive first**).

_____ 1

(b) Name the gas produced when metal **Y** reacts with cold water.

_____ 1

(c) Suggest names for metals **Y** and **Z**.

metal **Y** _____ metal **Z** _____ 1

(3)

[Turn over

Marks | KU | PS

16. The diagram shows the main stages in the making of malt whisky.

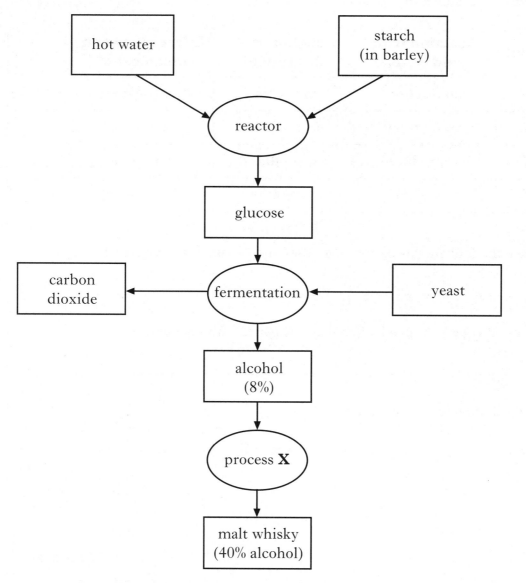

(a) Name the type of chemical reaction which takes place in the reactor.

_____

1

DO NOT WRITE IN THIS MARGIN

Marks KU PS

16. (continued)

(*b*) The equation for the reaction taking place during fermentation is:

$$C_6H_{12}O_6 \longrightarrow C_2H_5OH + CO_2$$

Balance this equation. **1**

(*c*) What name is given to process **X**?

______________________________ **1**

(*d*) Ethanol, C_2H_5OH, is the alcohol found in whisky.

A bottle of whisky contains 230 g of ethanol.

Calculate the number of moles of ethanol present in the whisky.

Show your working clearly.

__________ mol **2**

(5)

[Turn over

Marks | KU | PS

17. A student set up the cell shown.

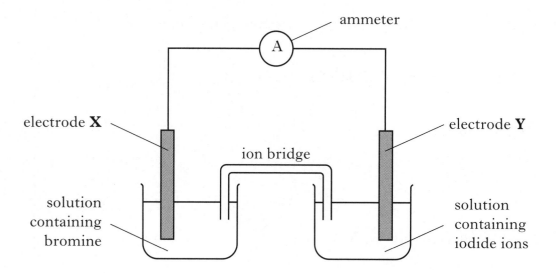

The reaction taking place at electrode **Y** is:

$$2I^-(aq) \longrightarrow I_2(s) + 2e^-$$

(a) Name the type of chemical reaction taking place at electrode **Y**.

_____   1

(b) **On the diagram**, clearly mark the path and direction of the electron flow.   1

(c) Describe a test, including the result, which would show that iodine had formed at electrode **Y**.

_____

_____   1

(d) Write the ion-electron equation for the chemical reaction taking place at electrode **X**.

1

(4)

DO NOT WRITE IN THIS MARGIN

Marks KU PS

18. When superglue sets, a polymer is formed.

Part of the polymer structure is shown.

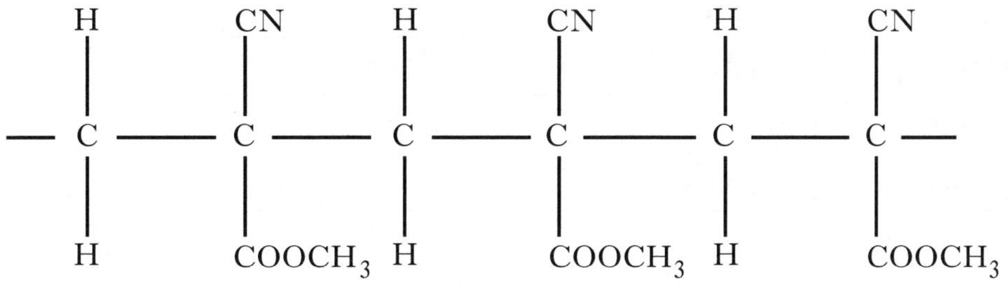

(*a*) Draw the structure of the repeating unit in the superglue polymer. **1**

(*b*) The polymer shown above contains methyl groups (CH_3).

Another type of superglue, used to close cuts, has the methyl groups replaced by either butyl groups (C_4H_9) or octyl groups.

Complete the table to show the number of carbon and hydrogen atoms in an octyl group.

	Number of atoms	
Group	**Carbon**	**Hydrogen**
methyl	1	3
butyl	4	9
octyl		

1

(*c*) Name a toxic gas given off when superglue burns.

____ **1**

(3)

[Turn over

Marks | KU | PS

19. (a) The table gives information about some members of the alkane family.

Name	Molecular formula	Boiling point/°C
nonane	C_9H_{20}	151
decane	$C_{10}H_{22}$	174
undecane	$C_{11}H_{24}$	196
dodecane	$C_{12}H_{26}$	

Predict the boiling point of dodecane.

_____ °C   1

(b) What term is used to describe any family of compounds, like the alkanes, which have the same general formula and similar chemical properties?

_____   1

(c) The equation for the burning of nonane is:

$$C_9H_{20} + 14O_2 \longrightarrow 9CO_2 + 10H_2O$$

Calculate the mass of water produced when 6·4 grams of nonane is burned.

Show your working clearly.

_____ g   2

Marks | KU | PS

19. (continued)

(*d*) Alkanes can be prepared by the Kolbé synthesis.

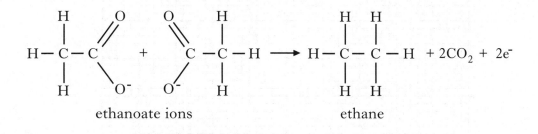

ethanoate ions   ethane

Draw a structural formula for the alkane produced when propanoate ions are used instead of ethanoate ions.

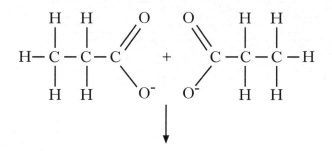

1

(5)

[END OF QUESTION PAPER]

ADDITIONAL SPACE FOR ANSWERS

ADDITIONAL GRAPH PAPER FOR QUESTION 13(*b*)(ii)

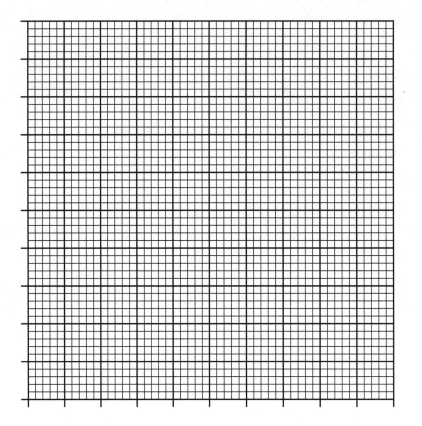

ADDITIONAL SPACE FOR ANSWERS

DO NOT
WRITE IN
THIS
MARGIN

KU	PS

ADDITIONAL SPACE FOR ANSWERS

STANDARD GRADE | CREDIT

2010

[BLANK PAGE]

FOR OFFICIAL USE

C

KU   PS

Total Marks

0500/402

NATIONAL
QUALIFICATIONS
2010

FRIDAY, 30 APRIL
10.50 AM – 12.20 PM

CHEMISTRY
STANDARD GRADE
Credit Level

Fill in these boxes and read what is printed below.

Full name of centre

Town

Forename

Surname

Date of birth

Day   Month   Year   Scottish candidate number   Number of seat

1 All questions should be attempted.

2 Necessary data will be found in the Data Booklet provided for Chemistry at Standard Grade and Intermediate 2.

3 The questions may be answered in any order but all answers are to be written in this answer book, and must be written clearly and legibly in ink.

4 Rough work, if any should be necessary, as well as the fair copy, is to be written in this book.

Rough work should be scored through when the fair copy has been written.

5 Additional space for answers and rough work will be found at the end of the book.

6 The size of the space provided for an answer should not be taken as an indication of how much to write. It is not necessary to use all the space.

7 Before leaving the examination room you must give this book to the Invigilator. If you do not, you may lose all the marks for this paper.

PART 1

In Questions 1 to 9 of this part of the paper, an answer is given by circling the appropriate letter (or letters) in the answer grid provided.

In some questions, two letters are required for full marks.

If more than the correct number of answers is given, marks will be deducted.

A total of 20 marks is available in this part of the paper.

SAMPLE QUESTION

A CH_4	B H_2	C CO_2
D CO	E C_2H_5OH	F C

(a) Identify the hydrocarbon.

Ⓐ	B	C
D	E	F

The one correct answer to part (a) is A. This should be circled.

(b) Identify the **two** elements.

A	Ⓑ	C
D	E	Ⓕ

As indicated in this question, there are **two** correct answers to part (b). These are B and F. Both answers are circled.

If, after you have recorded your answer, you decide that you have made an error and wish to make a change, you should cancel the original answer and circle the answer you now consider to be correct. Thus, in part (a), if you want to change an answer A to an answer D, your answer sheet would look like this:

Ⓐ̸	B	C
Ⓓ	E	F

If you want to change back to an answer which has already been scored out, you should enter a tick (✓) in the box of the answer of your choice, thus:

✓Ⓐ̸	B	C
Ⓓ̸	E	F

Marks | KU | PS

1. Crude oil can be separated into fractions.

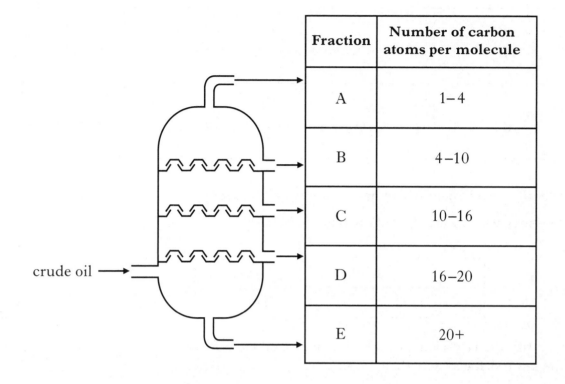

Fraction	Number of carbon atoms per molecule
A	1–4
B	4–10
C	10–16
D	16–20
E	20+

(*a*) Identify the fraction which is the most viscous.

A
B
C
D
E

1

(*b*) Identify the fraction used as camping gas.

A
B
C
D
E

1

(2)

[Turn over

Marks

2. The grid contains the symbols for some common elements.

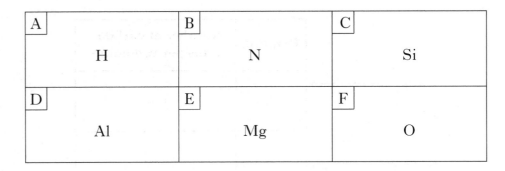

A H	B N	C Si
D Al	E Mg	F O

(a) Identify the element which has a density of $1.74\,g/cm^3$.

You may wish to use the data booklet to help you.

A	B	C
D	E	F

1

(b) Identify the **two** elements which react together to form a molecule with the same shape as a methane molecule.

A	B	C
D	E	F

1

(c) Identify the **two** elements which form an ionic compound with a formula of type X_2Y_3, where **X** is a metal.

A	B	C
D	E	F

1

(3)

Marks | KU | PS

3. The grid shows information about some particles.

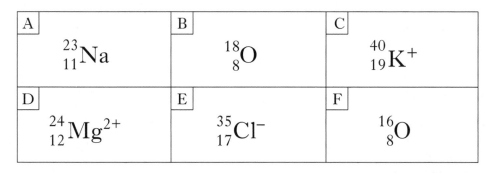

A	B	C
$^{23}_{11}Na$	$^{18}_{8}O$	$^{40}_{19}K^+$

D	E	F
$^{24}_{12}Mg^{2+}$	$^{35}_{17}Cl^-$	$^{16}_{8}O$

(*a*) Identify the **two** particles with the same number of neutrons.

A	B	C
D	E	F

1

(*b*) Identify the particle which has the same electron arrangement as neon.

A	B	C
D	E	F

1

(2)

[Turn over

DO NOT WRITE IN THIS MARGIN

Marks KU PS

4. The grid shows the names of some carbohydrates.

A	fructose
B	glucose
C	maltose
D	sucrose
E	starch

(a) Identify the condensation polymer.

| A |
| B |
| C |
| D |
| E |

1

(b) Identify the **two** monosaccharides.

| A |
| B |
| C |
| D |
| E |

1

(2)

DO NOT WRITE IN THIS MARGIN

Marks | KU | PS

5. Iron can be coated with different materials which provide a physical barrier against corrosion.

A	oil
B	zinc
C	plastic
D	tin
E	paint

(a) Identify the coating which is used to galvanise iron.

A
B
C
D
E

1

(b) Identify the coating which, if scratched, would cause the iron to rust faster than normal.

A
B
C
D
E

1

(2)

[Turn over

6. The structures of some hydrocarbons are shown in the grid.

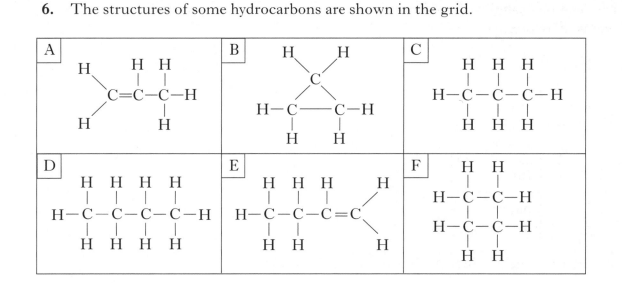

(a) Identify the **two** hydrocarbons with the general formula C_nH_{2n}, which do **not** react quickly with bromine solution.

A	B	C
D	E	F

1

(b) Identify the hydrocarbon which is the first member of a homologous series.

A	B	C
D	E	F

1

(c) Identify the **two** isomers of

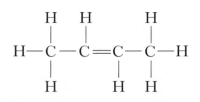

A	B	C
D	E	F

1

(3)

DO NOT WRITE IN THIS MARGIN

Marks KU PS

7. Elements can be used in different ways.

A chlorine	B potassium	C platinum
D hydrogen	E neon	F iron

(*a*) Identify the element which is a reactant in the Haber Process.

A	B	C
D	E	F

1

(*b*) Identify the element used as the catalyst in the manufacture of nitric acid (Ostwald Process).

A	B	C
D	E	F

1

(2)

[Turn over

Marks | KU | PS

8. The grid shows some statements which can be applied to different solutions.

A	It has a pH less than 7.
B	It conducts electricity.
C	It contains less $OH^-(aq)$ ions than pure water.
D	It does not neutralise dilute hydrochloric acid.
E	When diluted the concentration of $OH^-(aq)$ ions decreases.

Identify the **two** statements which are correct for an alkaline solution.

A
B
C
D
E

(2)

9. The grid shows pairs of chemicals.

A	B	C
copper carbonate + dilute sulphuric acid	lead nitrate solution + potassium iodide solution	potassium hydroxide + nitric acid
D	E	F
copper + water	silver + hydrochloric acid	ammonium nitrate + sodium hydroxide

Which **two** boxes contain a pair of chemicals that react together to form a gas?

A	B	C
D	E	F

(2)

[Turn over

Marks | KU | PS

PART 2

A total of 40 marks is available in this part of the paper.

10. Poly(methyl methacrylate) is a synthetic polymer used to manufacture perspex.

    (a) What is meant by the term **synthetic**?

    _____ 1

    (b) The structure of the methyl methacrylate monomer is shown.

$$
\begin{array}{cc}
H & CH_3 \\
| & | \\
C & = C \\
| & | \\
H & COOCH_3
\end{array}
$$

    methyl methacrylate

    (i) Draw a section of the poly(methyl methacrylate) polymer, showing three monomer units joined together.

    1

    (ii) Name the type of polymerisation taking place.

    _____ 1

    (c) Name a toxic gas produced when poly(methyl methacrylate) burns.

    _____ 1

    (4)

Marks | KU | PS

11. A student set up an experiment to investigate the breakdown of glucose to form alcohol.

At the start a deflated balloon was attached to the top of the tube.

After two hours the balloon inflates as shown.

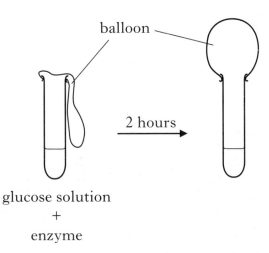

balloon

2 hours

glucose solution
+
enzyme

(a) (i) Name the type of chemical reaction taking place in the test tube.

_____ 1

(ii) Name the gas produced, which causes the balloon to inflate.

_____ 1

(b) The student repeated the experiment at 80 °C.

What effect would this have on how much the balloon inflates?

_____

_____

_____ 1

(3)

[Turn over

Marks KU PS

12. A student added magnesium ribbon to an excess of dilute sulphuric acid and measured the volume of hydrogen gas produced.

The reaction stopped when all the magnesium was used up.

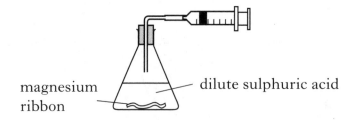

magnesium ribbon   dilute sulphuric acid

The results are shown in the table.

Time/s	0	10	20	40	50	60	70
Volume of hydrogen gas/cm^3	0	20	32	50	52	53	53

(a) State the test for hydrogen gas.

_____

_____   1

(b) Draw a line graph of the results.
Use appropriate scales to fill most of the graph paper.
(Additional graph paper, if required, will be found on page 24.)

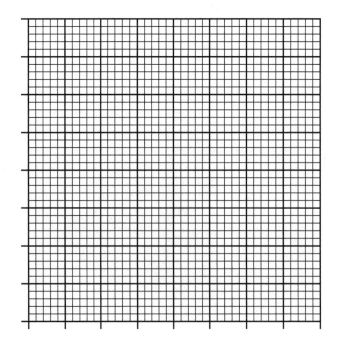

2

Marks

KU | PS

12. (continued)

(c) Using your graph, predict the volume of hydrogen gas produced during the first 30 seconds.

_____ cm^3 **1**

(d) The student repeated the experiment using a higher concentration of acid. The same volume of acid and the same mass of magnesium ribbon were used.

What volume of hydrogen gas would have been produced after 60 seconds?

_____ cm^3 **1**

(e) Calculate the mass of hydrogen produced when $4 \cdot 9$ g of magnesium reacts with an excess of dilute sulphuric acid.

$$Mg \ + \ H_2SO_4 \longrightarrow MgSO_4 \ + \ H_2$$

_____ g **2**

(7)

[Turn over

Marks

KU | PS

13. A student set up the following experiment to investigate the colour of ions in nickel(II) chromate solution.

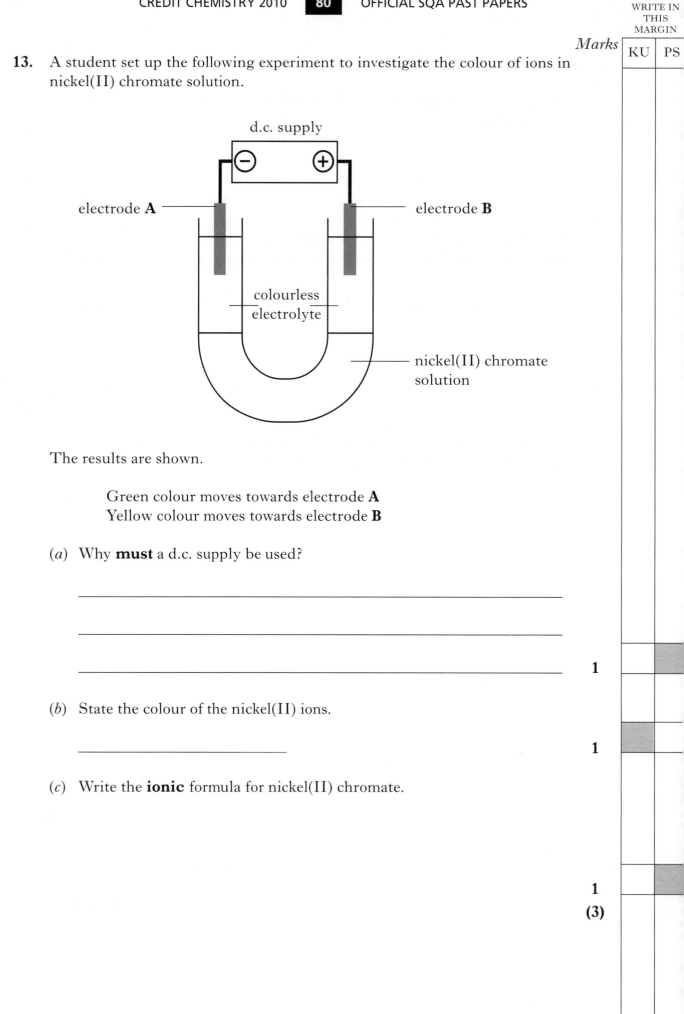

d.c. supply

electrode **A**

electrode **B**

colourless
electrolyte

nickel(II) chromate
solution

The results are shown.

Green colour moves towards electrode **A**
Yellow colour moves towards electrode **B**

(*a*) Why **must** a d.c. supply be used?

_____

_____

_____   **1**

(*b*) State the colour of the nickel(II) ions.

_____   **1**

(*c*) Write the **ionic** formula for nickel(II) chromate.

1

(3)

DO NOT
WRITE IN
THIS
MARGIN

Marks

KU | PS

14. The Eurofighter "Typhoon" is made from many newly developed materials including titanium alloys.

(*a*) The first step in extracting titanium from its ore is to convert it into titanium(IV) chloride.

Titanium(IV) chloride is a liquid at room temperature and does **not** conduct electricity.

What type of bonding, does this suggest, is present in titanium(IV) chloride?

_____

1

(*b*) Titanium(IV) chloride is then reduced to titanium metal.

The equation for the reaction taking place is:

$$TiCl_4 \quad + \quad Na \longrightarrow Ti \quad + \quad NaCl$$

(i) Balance the equation.

1

(ii) What does this reaction suggest about the reactivity of titanium compared to that of sodium?

_____

1

(3)

[Turn over

Marks KU PS

15. Scuba divers can suffer from painful and potentially fatal problems if they rise to the surface of the water too quickly. This causes dissolved nitrogen in their blood to form bubbles of nitrogen gas.

Distance from surface of water/m	Concentration of dissolved nitrogen/units
0	11·5
10	23·0
20	34·5
30	46·0
40	57·5

(a) Describe the relationship between the distance from the surface of the water and the concentration of dissolved nitrogen.

_____

_____

_____   1

(b) Predict the concentration of dissolved nitrogen at 60 m.

_____ units   1

(c) A nitrogen molecule is held together by three covalent bonds.

Circle the correct words to complete the sentence.

In a covalent bond the atoms are held together by the attraction

between the positive { electrons / neutrons / protons } and the shared pair of negative

{ electrons / neutrons / protons }.   1

(3)

DO NOT
WRITE IN
THIS
MARGIN

Marks | KU | PS

16. (*a*) Galena is an ore containing lead sulphide, PbS.

(i) What is the charge on this lead ion?

_____

1

(ii) Calculate the percentage by mass of lead in galena, PbS.

_____ %   2

(*b*) Most metals have to be extracted from their ores.

(i) Name the metal extracted in a Blast furnace.

_____

1

(ii) Place the following metals in the correct space in the table.

copper, mercury, aluminium

You may wish to use the data booklet to help you.

Metal	Method of extraction
	using heat alone
	electrolysis of molten ore
	heating with carbon

1

(5)

[**Turn over**

Marks | KU | PS

17. Iron displaces silver from silver(I) nitrate solution.

The equation for the reaction is:

$$Fe(s) + 2Ag^+(aq) + 2NO_3^-(aq) \longrightarrow Fe^{2+}(aq) + 2Ag(s) + 2NO_3^-(aq)$$

(a) (Circle) the spectator ion in the above equation.   **1**

(b) Describe a chemical test, including the result, to show that $Fe^{2+}(aq)$ ions are formed.

_____

_____   **1**

(c) Write the ion-electron equation for the **reduction** step in the reaction.

You may wish to use the data book to help you.

1

(d) This reaction can also be carried out in a cell.

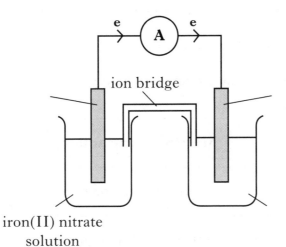

iron(II) nitrate
solution

Complete the three labels on the diagram.   **1**

(An additional diagram, if required, will be found on page 24.)   **(4)**

Marks

18. Ethers are useful chemicals.

Some are listed in the table.

Structural formula	Name of ether
$CH_3CH_2 - O - CH_2CH_3$	ethoxyethane
$CH_3 - O - CH_2CH_2CH_3$	methoxypropane
$CH_3 - O - CH_2CH_3$	methoxyethane
$CH_3CH_2 - O - CH_2CH_2CH_3$	**X**

(a) Suggest a name for ether **X**.

_____

1

(b) The boiling points of ethers and alkanes are approximately the same when they have a **similar** relative formula mass.

Suggest the **boiling point** of ethoxyethane (relative formula mass 74).

You may wish to use the data booklet to help you.

_____ °C   1

(2)

[Turn over

Marks | KU | PS

19. A student carried out a titration using the chemicals and apparatus below.

hydrochloric acid
0·1 mol/l

	Rough titre	1st titre	2nd titre
Initial burette reading/cm^3	0·3	0·2	0·5
Final burette reading/cm^3	26·6	25·3	25·4
Volume used/cm^3	26·3	25·1	24·9

10 cm^3
sodium hydroxide
solution + indicator

(a) Using the results in the table, calculate the **average** volume of hydrochloric acid required to neutralise the sodium hydroxide solution.

_____ cm^3 **1**

(b) The equation for the reaction is:

$$HCl \ + \ NaOH \longrightarrow H_2O \ + \ NaCl$$

Using your answer from part (a), calculate the concentration of the sodium hydroxide solution.

Show your working clearly.

_____ mol/l   **2**

(3)

Marks | KU | PS

20. Chemists have discovered a way to insert a –CH_2– group into any bond which includes an atom of hydrogen.

When a –CH_2– group is inserted into a methanol molecule the following reaction takes place.

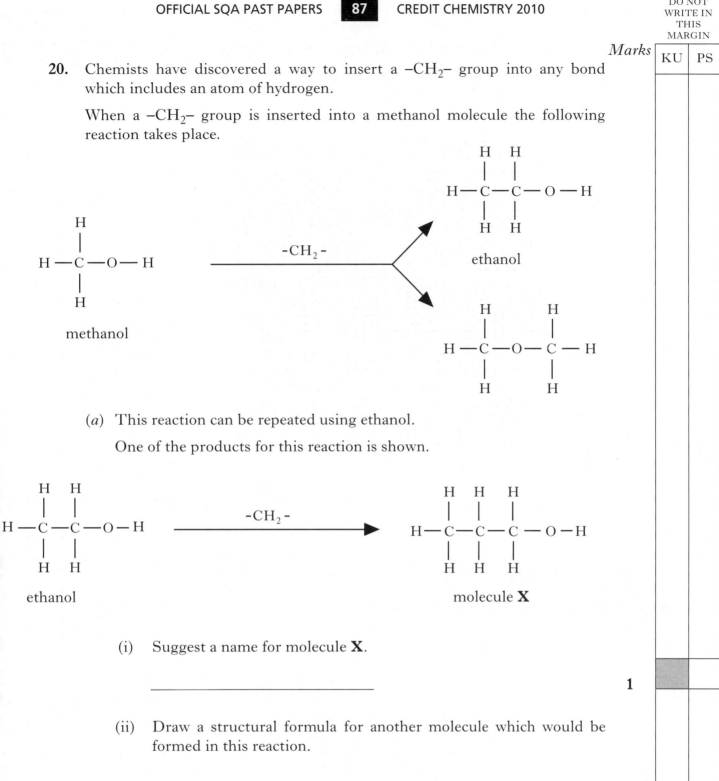

methanol

ethanol

(a) This reaction can be repeated using ethanol.

One of the products for this reaction is shown.

ethanol

molecule **X**

(i) Suggest a name for molecule **X**.

_____

1

(ii) Draw a structural formula for another molecule which would be formed in this reaction.

1

(b) Identify the **two** products formed when molecule **X** is completely burned in a plentiful supply of oxygen.

_____

1

(3)

[END OF QUESTION PAPER]

ADDITIONAL SPACE FOR ANSWERS

ADDITIONAL GRAPH PAPER FOR QUESTION 12(*b*)

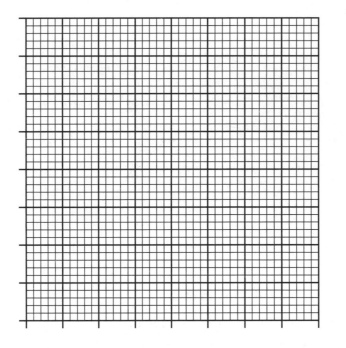

ADDITIONAL DIAGRAM FOR QUESTION 17(*d*)

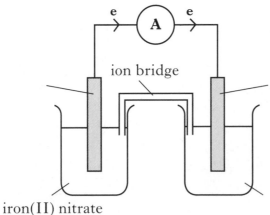

iron(II) nitrate solution

[BLANK PAGE]

C

FOR OFFICIAL USE

	KU	PS
Total Marks		

0500/402

NATIONAL
QUALIFICATIONS
2011

THURSDAY, 26 MAY
10.50 AM – 12.20 PM

**CHEMISTRY
STANDARD GRADE**
Credit Level

Fill in these boxes and read what is printed below.

Full name of centre

Town

Forename(s)

Surname

Date of birth

Day	Month	Year	Scottish candidate number	Number of seat

1 All questions should be attempted.

2 Necessary data will be found in the Data Booklet provided for Chemistry at Standard Grade and Intermediate 2.

3 The questions may be answered in any order but all answers are to be written in this answer book, and must be written clearly and legibly in ink.

4 Rough work, if any should be necessary, as well as the fair copy, is to be written in this book.

Rough work should be scored through when the fair copy has been written.

5 Additional space for answers and rough work will be found at the end of the book.

6 The size of the space provided for an answer should not be taken as an indication of how much to write. It is not necessary to use all the space.

7 Before leaving the examination room you must give this book to the Invigilator. If you do not, you may lose all the marks for this paper.

PART 1

In Questions 1 to 10 of this part of the paper, an answer is given by circling the appropriate letter (or letters) in the answer grid provided.

In some questions, two letters are required for full marks.

If more than the correct number of answers is given, marks will be deducted.

A total of 20 marks is available in this part of the paper.

SAMPLE QUESTION

A CH_4	B H_2	C CO_2
D CO	E C_2H_5OH	F C

(a) Identify the hydrocarbon.

Ⓐ	B	C
D	E	F

The one correct answer to part (a) is A. This should be circled.

(b) Identify the **two** elements.

A	Ⓑ	C
D	E	Ⓕ

As indicated in this question, there are **two** correct answers to part (b). These are B and F. Both answers are circled.

If, after you have recorded your answer, you decide that you have made an error and wish to make a change, you should cancel the original answer and circle the answer you now consider to be correct. Thus, in part (a), if you want to change an answer A to an answer D, your answer sheet would look like this:

A̶	B	C
Ⓓ	E	F

If you want to change back to an answer which has already been scored out, you should enter a tick (✓) in the box of the answer of your choice, thus:

✓A̶	B	C
D̶	E	F

DO NOT WRITE IN THIS MARGIN

Marks KU PS

1. Limewater can be made by dissolving calcium hydroxide in water.

Identify the term used to describe the water.

A	solute
B	solvent
C	solution
D	insoluble

A
B
C
D

(1)

[Turn over

Marks | KU | PS

2. Distillation of crude oil produces several fractions.

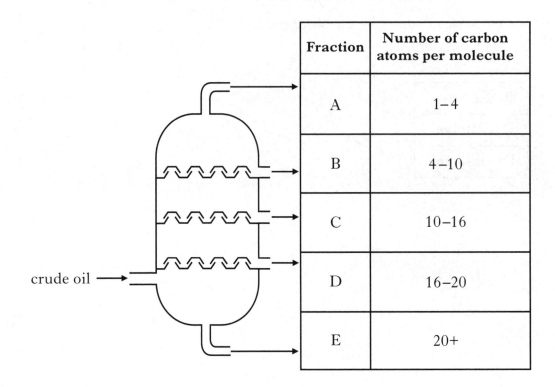

Fraction	Number of carbon atoms per molecule
A	1–4
B	4–10
C	10–16
D	16–20
E	20+

crude oil →

(a) Identify the fraction which is used to tar roads.

A
B
C
D
E

1

(b) Identify the fraction which is most flammable.

A
B
C
D
E

1

(2)

Marks　KU　PS

3. The grid shows the symbols of some elements.

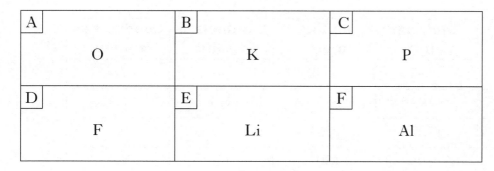

A		B		C	
	O		K		P
D		E		F	
	F		Li		Al

(a) Identify the element with the lowest density.

You may wish to use the data booklet to help you.

A	B	C
D	E	F

1

(b) Identify the **two** elements which can form ions with the same electron arrangement as argon.

You may wish to use the data booklet to help you.

A	B	C
D	E	F

1

(c) Identify the **two** elements which would react together to form a molecule with the same shape as an ammonia molecule.

A	B	C
D	E	F

1

(3)

[Turn over

Marks | KU | PS

4. The table contains information about some substances.

Substance	Melting point/°C	Boiling point/°C	Conducts as a solid	Conducts as a liquid
A	−7	59	no	no
B	1492	2897	yes	yes
C	1407	2357	no	no
D	606	1305	no	yes
E	−39	357	yes	yes
F	−78	−33	no	no

(a) Identify the substance which is a gas at 0 °C.

A
B
C
D
E
F

1

(b) Identify the **two** substances which exist as molecules.

A
B
C
D
E
F

1

(2)

Marks | KU | PS

5. The grid shows the formulae of some oxides.

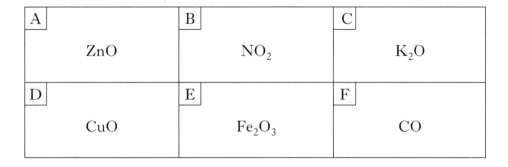

A	B	C
ZnO	NO$_2$	K$_2$O
D	E	F
CuO	Fe$_2$O$_3$	CO

(a) Identify the **two** oxides which are covalent.

A	B	C
D	E	F

1

(b) Identify the oxide which dissolves in water to give an alkaline solution.

You may wish to use the data booklet to help you.

A	B	C
D	E	F

1

(c) Identify the oxide which is reduced in a blast furnace.

A	B	C
D	E	F

1

(3)

[Turn over

Marks | KU | PS

6. Equations are used to represent chemical reactions.

A	$2H_2(g) + O_2(g) \longrightarrow 2H_2O(\ell)$
B	$2H_2O(\ell) + O_2(g) + 4e^- \longrightarrow 4OH^-(aq)$
C	$CH_4(g) + 2O_2(g) \longrightarrow CO_2(g) + 2H_2O(\ell)$
D	$H^+(aq) + OH^-(aq) \longrightarrow H_2O(\ell)$
E	$Zn(s) + FeSO_4(aq) \longrightarrow Fe(s) + ZnSO_4(aq)$

(a) Identify the equation which represents neutralisation.

| A |
| B |
| C |
| D |
| E |

1

(b) Identify the equation involved in the rusting of iron.

| A |
| B |
| C |
| D |
| E |

1

(2)

Marks

	KU	PS

7. A student made the following statements about the particles found in an atom.

A	Relative mass = 1
B	Charge = zero
C	Found outside the nucleus
D	Charge = 1+
E	Charge = 1–

Identify the **two** statements which apply to an electron.

A
B
C
D
E

(1)

[Turn over

Marks   KU   PS

8.   Identify the **two** statements which apply to zinc.

You may wish to use the data booklet to help you.

A	It displaces calcium from a solution of calcium nitrate.
B	It reacts with cold water.
C	It can be obtained by heating its oxide.
D	It reacts with dilute hydrochloric acid.
E	It is displaced from a solution of its chloride by magnesium.

A
B
C
D
E

(2)

9. The diagram shows how an object can be coated with silver.

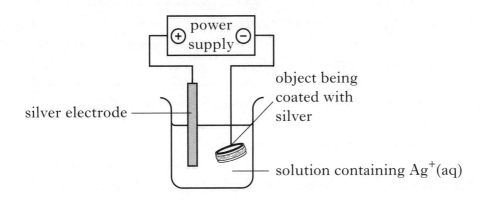

The following reactions take place at the electrodes.

Negative electrode: $Ag^+(aq) + e^- \longrightarrow Ag(s)$

Positive electrode: $Ag(s) \longrightarrow Ag^+(aq) + e^-$

Identify the **two** correct statements.

A	Ions flow through the solution.
B	Silver ions move towards the silver electrode.
C	The process is an example of galvanising.
D	The mass of the silver electrode decreases.
E	Reduction occurs at the silver electrode.

A
B
C
D
E

(2)

[Turn over

Marks KU PS

10. A student made the following statements about the rusting of iron.

A	During rusting Fe^{3+} ions are changed to Fe^{2+} ions.
B	Rusting is an example of oxidation.
C	Iron rusts when connected to the negative terminal of a battery.
D	Tin gives sacrificial protection to iron.
E	Electroplating provides a surface barrier to air and water.

Identify the **two** correct statements.

A
B
C
D
E

(2)

[Turn over for Part 2 on *Page fourteen*

Marks | KU | PS

PART 2

A total of 40 marks is available in this part of the paper.

11. (*a*) The table shows information about two of the gases found in air.

Gas	Boiling point/°C
oxygen	−183
nitrogen	−196

At very low temperatures air is a mixture of liquids.

Name the process which can be used to separate this mixture.

_____   1

(*b*) In a sample of oxygen there are two different types of oxygen atom:

$^{18}_{8}O$ and $^{16}_{8}O$

(i) What **term** is used to describe these different types of oxygen atom?

_____   1

(ii) Complete the table for each type of oxygen atom.

Type of atom	Number of protons	Number of neutrons
$^{18}_{8}O$		
$^{16}_{8}O$		

1

(3)

12. (a) Ethanol, for alcoholic drinks, can be made from glucose.

Name this process.

_____

1

(b) The table below shows the relationship between the percentage of ethanol and the density of alcoholic drinks.

Percentage of ethanol (%)	40	50	60	70	80
Density of alcoholic drink (g/cm^3)	0·928	0·907	0·886	0·865	0·844

(i) Write a general statement describing how the percentage of ethanol affects the density of the alcoholic drink.

_____

_____

_____

1

(ii) The density of a particular brand of alcoholic drink is $0·970\,g/cm^3$.

Predict the percentage of ethanol in this alcoholic drink.

_____ % 1

(3)

[Turn over

Marks | KU | PS

13. Polyvinyldichloride (PVDC) is a plastic used in food packaging.

The structure of part of a PVDC molecule is shown.

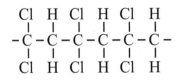

(*a*) Draw the **full** structural formula for the monomer used to make PVDC.

1

(*b*) Name a toxic gas produced when PVDC burns.

_____

1

(2)

Marks | KU | PS

14. (*a*) When sulphur dioxide dissolves in water in the atmosphere "acid rain" is produced.

Circle the correct phrase to complete the sentence.

Compared with pure water, acid rain contains $\begin{cases} \text{a higher} \\ \text{a lower} \\ \text{the same} \end{cases}$ concentration of hydrogen ions. **1**

(*b*) The table shows information about the solubility of sulphur dioxide.

Temperature /°C	0	20	30	40	50	60
Solubility in g/100 cm³	22·0	10·0	6·0	3·0	2·0	1·5

Draw a line graph of solubility against temperature.

Use appropriate scales to fill most of the graph paper.

(Additional graph paper, if required, will be found on page 28.)

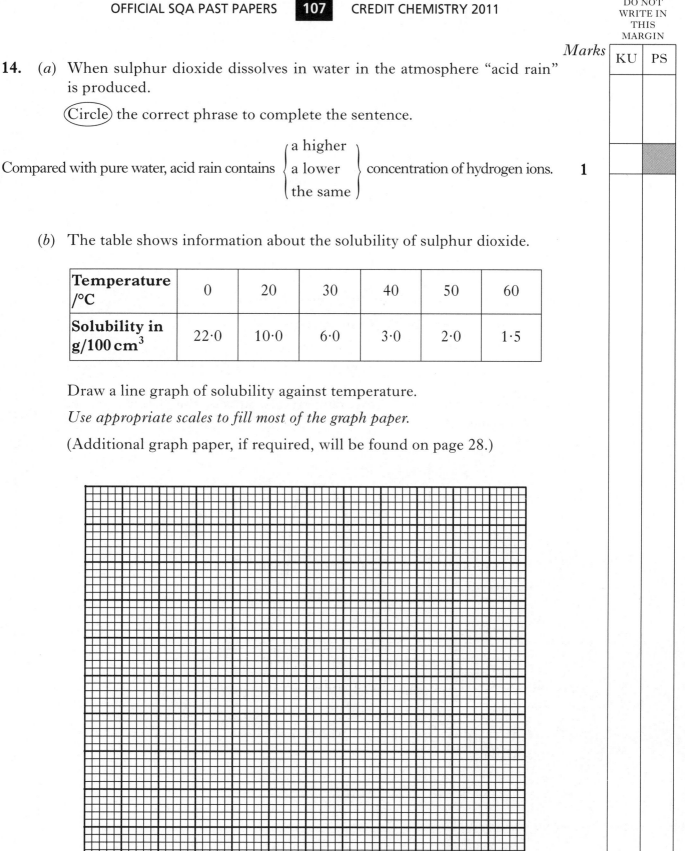

2

(3)

Marks | KU | PS

15. Scientists have developed a "bio-battery" which produces electricity from sucrose.

(*a*) Write the molecular formula for sucrose.

1

(*b*) Name an isomer of sucrose.

_____

1

(*c*) The sucrose is broken down using an enzyme.

(i) What is meant by the term "enzyme"?

_____

1

(ii) The graph shows how temperature affects the activity of an enzyme.

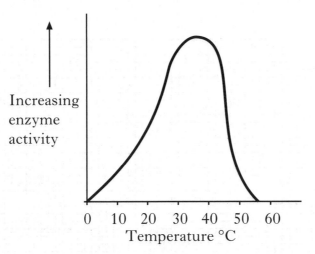

State **one** other factor which has a similar effect on enzyme activity.

_____

1

(4)

Marks   KU | PS

16. Heptane can be cracked as shown.

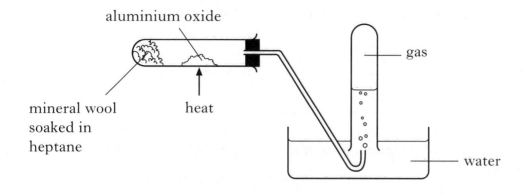

One of the reactions which takes place is:

$$C_7H_{16} \longrightarrow C_4H_{10} + C_3H_6$$

(*a*) The product C_3H_6 decolourises bromine solution quickly.

Draw a structural formula for an isomer of C_3H_6, which would **not** decolourise bromine solution quickly.

1

(*b*) Aluminium oxide is used as a catalyst to speed up the reaction.

(i) Suggest another reason for using a catalyst.

_____

_____

1

(ii) Write the formula for aluminium oxide.

1

(3)

[Turn over

DO NOT WRITE IN THIS MARGIN

Marks KU PS

17. Urea reacts with water, breaking down to form carbon dioxide and ammonia.

$$\underset{\text{urea}}{H_2NCONH_2} + H_2O \longrightarrow CO_2 + 2NH_3$$

(*a*) Suggest a name for the **type** of chemical reaction taking place.

____ **1**

(*b*) Calculate the mass of ammonia produced, in grams, when 90 g of urea breaks down.

____________ g **2**

(3)

DO NOT
WRITE IN
THIS
MARGIN

Marks | KU | PS

18. A student set up the following experiment to electrolyse cobalt chloride solution.

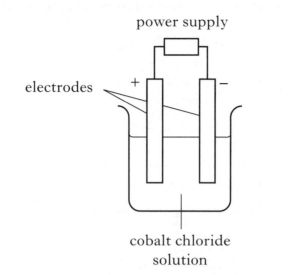

power supply

electrodes

+   −

cobalt chloride
solution

(*a*) What **type** of power supply **must** be used to electrolyse cobalt chloride solution?

_____   1

(*b*) Describe what would be **seen** at the positive electrode.

You may wish to use the data booklet to help you.

_____

_____   1

(*c*) The formula for cobalt chloride is $CoCl_2$.

What is the charge on the cobalt ion in $CoCl_2$?

_____   1

(3)

[Turn over

Marks   KU   PS

19. Catalysts can be used in different processes.

(*a*) The flow diagram shows the steps involved in the Haber process.

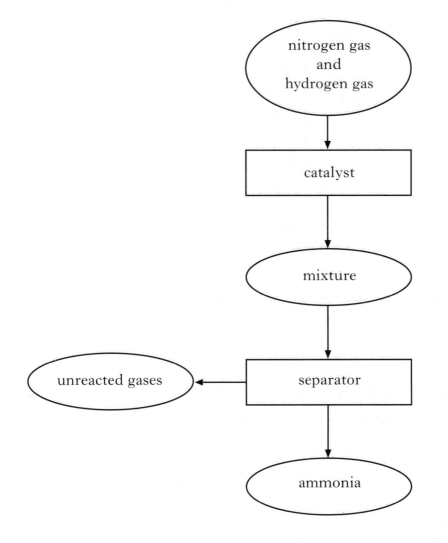

On the flow diagram above draw an arrow to show how the process is made more economical.

1

Marks | KU | PS

19. **(continued)**

(b) Ammonia can be used to produce nitrogen dioxide as shown.

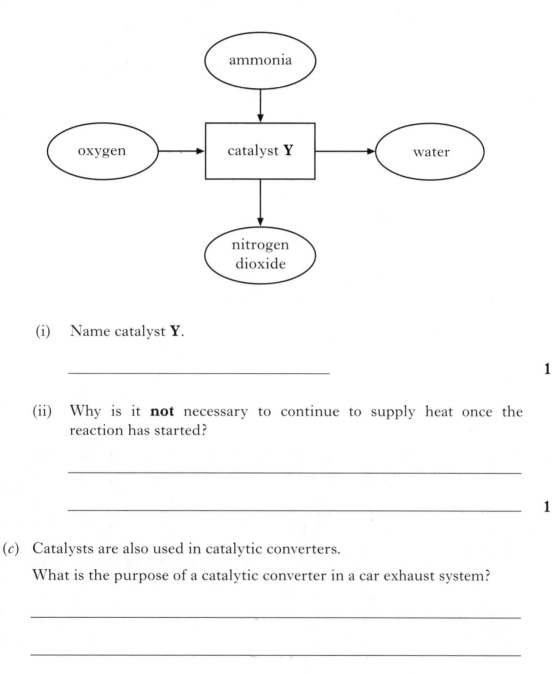

(i) Name catalyst **Y**.

_____

1

(ii) Why is it **not** necessary to continue to supply heat once the reaction has started?

_____

_____

1

(c) Catalysts are also used in catalytic converters.

What is the purpose of a catalytic converter in a car exhaust system?

_____

_____

_____

1

(4)

[Turn over

Marks | KU | PS

20. Metal salts can be produced by different methods.

(a) Lead(II) iodide can be produced by reacting lead(II) nitrate solution with sodium iodide solution.

The equation for this reaction is:

$$Pb(NO_3)_2(aq) \ + \ NaI(aq) \longrightarrow PbI_2(s) \ + \ NaNO_3(aq)$$

(i) Balance the above equation.

1

(ii) What technique could be used to remove lead(II) iodide from the mixture?

_____

1

(b) The salt copper(II) nitrate can be produced as shown.

$$\mathbf{X} \ + \ 2HNO_3 \longrightarrow Cu(NO_3)_2 \ + \ CO_2 \ + \ H_2O$$

Name substance **X**.

_____

1

(c) Potassium sulphate can be produced by titrating potassium hydroxide solution with dilute sulphuric acid.

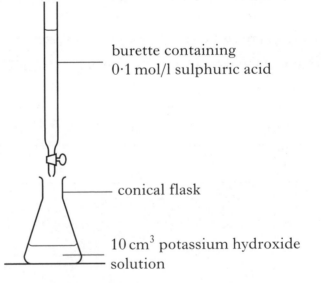

burette containing
0·1 mol/l sulphuric acid

conical flask

10 cm³ potassium hydroxide solution

Marks | KU | PS

20. *(c)* **(continued)**

(i) What must be added to the conical flask to show the end-point of the titration?

_____   1

(ii) The average volume of sulphuric acid used in the titration is $20 \, cm^3$.

Calculate the number of moles of sulphuric acid used.

_____ mol   1

(d) The equation for the reaction is:

$$H_2SO_4 \; + \; 2KOH \longrightarrow K_2SO_4 \; + \; 2H_2O$$

Using your answer from part *(c)*(ii), calculate the number of moles of potassium hydroxide in the $10 \, cm^3$ sample of potassium hydroxide solution.

_____ mol   1

(6)

[Turn over

Marks | KU | PS

21. A technician set up the following cell.

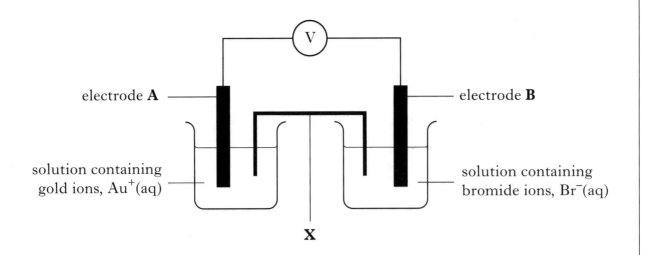

electrode **A**

electrode **B**

solution containing
gold ions, Au$^+$(aq)

solution containing
bromide ions, Br$^-$(aq)

X

The reaction taking place at electrode **B** is:

$$2Br^-(aq) \longrightarrow Br_2(\ell) + 2e^-$$

(a) **On the diagram**, clearly mark the path and direction of electron flow.   1

(b) Write the ion-electron equation for the reaction taking place at electrode **A**.

You may wish to use the data booklet to help you.

1

(c) Name the piece of apparatus labelled **X**.

_____   1

(3)

Marks   KU   PS

22. Ethylthioethane belongs to a homologous series of compounds called thioethers.

(a) What is meant by a homologous series?

_____

_____

_____   1

(b) Ethylthioethane is formed when ethylthiol reacts with bromoethane as shown.

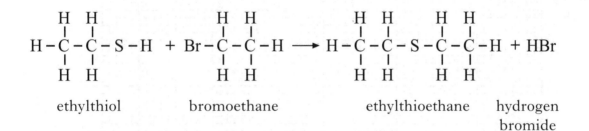

ethylthiol   bromoethane   ethylthioethane   hydrogen bromide

Draw the **full** structural formula for the thioether produced in the following reaction.

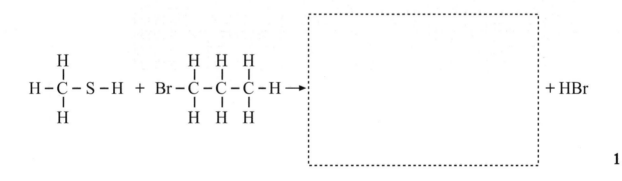

1

(c) Ethylthioethane can also be formed by the reaction of ethylthiol with ethene.

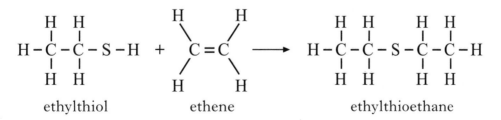

ethylthiol   ethene   ethylthioethane

Suggest a name for the **type** of chemical reaction taking place.

_____   1

(3)

[*END OF QUESTION PAPER*]

ADDITIONAL SPACE FOR ANSWERS

ADDITIONAL GRAPH PAPER FOR QUESTION 14(*b*)

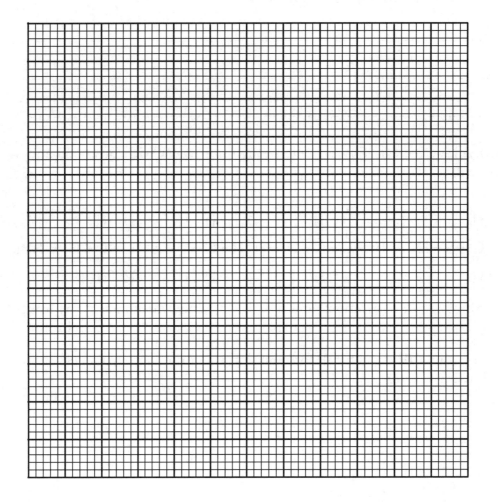

ADDITIONAL SPACE FOR ANSWERS

ADDITIONAL SPACE FOR ANSWERS

[BLANK PAGE]

FOR OFFICIAL USE

C

KU   PS

Total
Marks

0500/31/01

NATIONAL
QUALIFICATIONS
2012

MONDAY, 14 MAY
10.50 AM – 12.20 PM

CHEMISTRY
STANDARD GRADE
Credit Level

Fill in these boxes and read what is printed below.

Full name of centre

Town

Forename(s)

Surname

Date of birth

Day   Month   Year   Scottish candidate number   Number of seat

1   All questions should be attempted.

2   Necessary data will be found in the Data Booklet provided for Chemistry at Standard Grade and Intermediate 2.

3   The questions may be answered in any order but all answers are to be written in this answer book, and must be written clearly and legibly in ink.

4   Rough work, if any should be necessary, as well as the fair copy, is to be written in this book.
   Rough work should be scored through when the fair copy has been written.

5   Additional space for answers and rough work will be found at the end of the book.

6   The size of the space provided for an answer should not be taken as an indication of how much to write.   It is not necessary to use all the space.

7   Before leaving the examination room you must give this book to the Invigilator.   If you do not, you may lose all the marks for this paper.

PART 1

In Questions 1 to 9 of this part of the paper, an answer is given by circling the appropriate letter (or letters) in the answer grid provided.

In some questions, two letters are required for full marks.

If more than the correct number of answers is given, marks will be deducted.

A total of 20 marks is available in this part of the paper.

SAMPLE QUESTION

A CH_4	B H_2	C CO_2
D CO	E C_2H_5OH	F C

(a)   Identify the hydrocarbon.

Ⓐ	B	C
D	E	F

The one correct answer to part (a) is A. This should be circled.

(b)   Identify the **two** elements.

A	Ⓑ	C
D	E	Ⓕ

As indicated in this question, there are **two** correct answers to part (b). These are B and F.

Both answers are circled.

If, after you have recorded your answer, you decide that you have made an error and wish to make a change, you should cancel the original answer and circle the answer you now consider to be correct. Thus, in part (a), if you want to change an answer A to an answer D, your answer sheet would look like this:

Ⱥ̸	B	C
Ⓓ	E	F

If you want to change back to an answer which has already been scored out, you should enter a tick (✓) in the box of the answer of your choice, thus:

✓Ⱥ̸	B	C
Ⱦ̸	E	F

Marks | KU | PS

1. The grid shows the formulae of some gases.

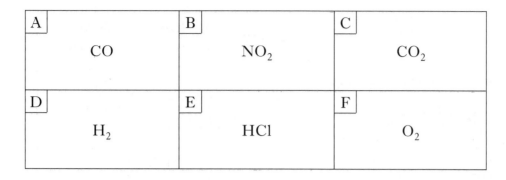

A	B	C
CO	NO_2	CO_2
D	E	F
H_2	HCl	O_2

(a) Identify the **two** toxic gases produced during the burning of polyvinylchloride (PVC).

A	B	C
D	E	F

1

(b) Identify the gas which burns with a pop.

A	B	C
D	E	F

1

(2)

[Turn over

Marks   KU   PS

2. The table shows some fractions from crude oil.

Fraction	Boiling range/°C	Name of fraction
A	−160 to 20 °C	Refinery Gas
B	20 to 120 °C	Naphtha
C	120 to 240 °C	Kerosene
D	240 to 350 °C	Gas Oils
E	Over 350 °C	Residue

Crude oil →

(*a*) Identify the fraction with the longest chain length.

A
B
C
D
E

1

(*b*) Identify the fraction which is used as fuel for aeroplanes.

A
B
C
D
E

1

(2)

Marks

KU | PS

3. Lead(II) nitrate solution reacts with potassium iodide solution to give a yellow solid.

$$Pb^{2+}(aq) + 2NO_3^{-}(aq) + 2K^{+}(aq) + 2I^{-}(aq) \rightarrow Pb^{2+}(I^{-})_2(s) + 2K^{+}(aq) + 2NO_3^{-}(aq)$$

Identify the **two** spectator ions in the reaction.

A	Pb^{2+}
B	NO_3^{-}
C	K^{+}
D	I^{-}

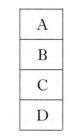

(1)

[Turn over

Marks | KU | PS

4. The grid shows information about some particles.

| Particle | Number of | | |
	protons	neutrons	electrons
A	11	12	11
B	9	10	9
C	11	13	11
D	19	20	18
E	9	10	10

(a) Identify the particle which is a negative ion.

A
B
C
D
E

1

(b) Identify the particle which would give a lilac flame colour.

You may wish to use the data booklet to help you.

A
B
C
D
E

1

(c) Identify the **two** particles which are isotopes.

A
B
C
D
E

1

(3)

Marks | KU | PS

5. The grid shows the structural formulae of some hydrocarbons.

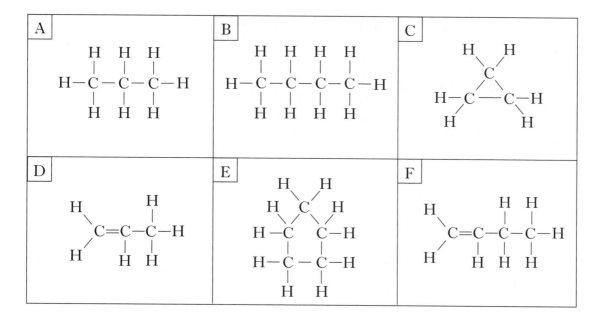

(a) Identify the hydrocarbon which reacts with hydrogen to form butane.

A	B	C
D	E	F

1

(b) Identify the **two** isomers.

A	B	C
D	E	F

1

(c) Identify the structural formula which represents propene.

A	B	C
D	E	F

1

(3)

[Turn over

Marks | KU | PS

6. Equations are used to represent chemical reactions.

A	$Zn(s) \longrightarrow Zn^{2+}(aq) + 2e^-$
B	$C_2H_5OH(\ell) + 3O_2(g) \longrightarrow 2CO_2(g) + 3H_2O(\ell)$
C	$SO_2(g) + H_2O(\ell) \longrightarrow 2H^+(aq) + SO_3^{2-}(aq)$
D	$H^+(aq) + OH^-(aq) \longrightarrow H_2O(\ell)$
E	$SO_4^{2-}(aq) + 2H^+(aq) + 2e^- \longrightarrow SO_3^{2-}(aq) + H_2O(\ell)$

(a) Identify the equation which represents the formation of acid rain.

A
B
C
D
E

1

(b) Identify the equation which represents combustion.

A
B
C
D
E

1

(2)

7. The grid shows the names of some soluble compounds.

A	B	C
sodium iodide	potassium chloride	lithium chloride
D	E	F
barium bromide	sodium hydroxide	potassium sulphate

(a) Identify the base.

A	B	C
D	E	F

1

(b) Identify the **two** compounds whose solutions would form a precipitate when mixed.

You may wish to use the data booklet to help you.

A	B	C
D	E	F

1

(c) Identify the compound with a formula of the type XY_2, where X is a metal.

A	B	C
D	E	F

1

(3)

[Turn over

8. The grid shows the names of some processes.

A	distillation
B	precipitation
C	filtering
D	electrolysis
E	dissolving

(a) Identify the process which is used to increase the alcohol concentration of fermentation products.

A
B
C
D
E

1

(b) Identify the **two** processes which should be used to separate magnesium carbonate from a mixture of solid magnesium carbonate and solid magnesium chloride.

You may wish to use the data booklet to help you.

A
B
C
D
E

1

(2)

Marks | KU | PS

9. A student made some statements about the particles in an atom.

A	It has a negative charge.
B	It is found inside the nucleus.
C	It has zero charge.
D	It is found outside the nucleus.
E	It has a relative mass of almost zero.
F	It has a relative mass of 1.

Identify the **two** statements which apply to a proton.

A
B
C
D
E
F

(2)

[Turn over for Part 2 on *Page twelve*

Marks KU PS

PART 2

A total of 40 marks is available in this part of the paper.

10. Iron can be coated with a physical barrier to prevent rusting.

    (a) How does coating iron prevent rusting?

    _____

    _____   1

    (b) A student investigated the rusting of iron. The coatings on four strips
    of iron were **scratched** to expose the iron. The strips were then placed
    in salt water.

    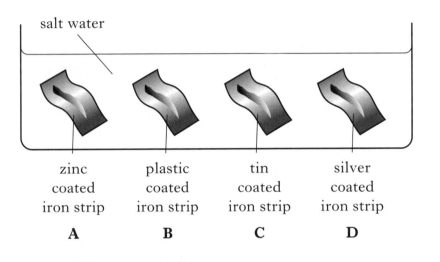

zinc coated iron strip	plastic coated iron strip	tin coated iron strip	silver coated iron strip
A	**B**	**C**	**D**

    (i) Which iron strip has been galvanised, **A**, **B**, **C** or **D**?

    _____   1

    (ii) Which iron strip would have rusted most quickly, **A**, **B**, **C** or **D**?

    _____   1

    (3)

Marks KU PS

11. A student carried out some experiments between zinc and excess 1 mol/l hydrochloric acid.

The graph shows the results of each experiment.

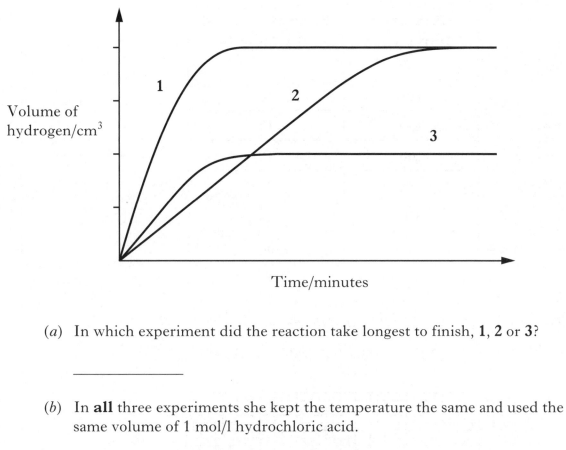

(a) In which experiment did the reaction take longest to finish, 1, 2 or 3?

_____

1

(b) In **all** three experiments she kept the temperature the same and used the same volume of 1 mol/l hydrochloric acid.

(i) Suggest one factor that could have been changed from experiment 1 to produce the results in experiment 2.

_____

1

(ii) 1 g of zinc was used in experiment 1.

What mass of zinc was used in experiment 3?

_____ g   1

(3)

[Turn over

Marks | KU | PS

12. Ammonia is produced in the Haber process.

The percentage yield of ammonia, obtained at different pressures, is shown in the table.

Pressure/ atmospheres	Percentage yield of ammonia
50	6
100	10
150	14
200	19
250	22
350	29
400	32

(a) Draw a line graph of the results.

Use appropriate scales to fill most of the graph paper.

(Additional graph paper, if required, can be found on page 26.)

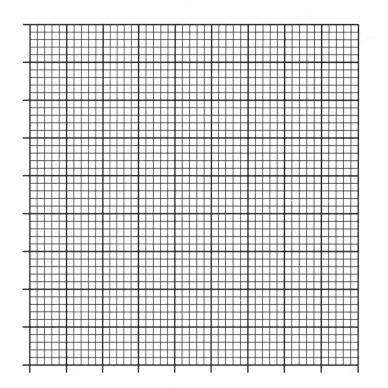

2

DO NOT WRITE IN THIS MARGIN

Marks KU PS

12. (continued)

(*b*) Using your graph, estimate the yield of ammonia at 300 atmospheres.

____________ % **1**

(*c*) Temperature is another factor which affects the percentage yield of ammonia.

Temperature/°C	Percentage yield of ammonia
200	88
300	67
400	49
500	18

Suggest a reason why 500 °C is the temperature chosen to operate an industrial ammonia plant rather than 200 °C.

____

____ **1**

(4)

[Turn over

Marks KU PS

13. Hydrogen gas is made up of diatomic molecules.

(a) Draw a diagram to show how the electrons are arranged in a molecule of hydrogen, H_2.

1

(b) Hydrogen gas is produced when magnesium reacts with dilute sulphuric acid.

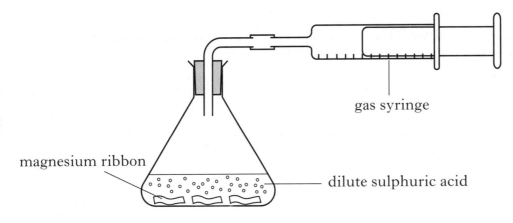

gas syringe

magnesium ribbon

dilute sulphuric acid

The equation for the reaction is:

$$Mg(s) \ + \ H_2SO_4(aq) \longrightarrow MgSO_4(aq) \ + \ H_2(g)$$

(i) (Circle) the formula for the salt in the above equation. 1

Marks | KU | PS

13. (*b*) (continued)

   (ii) The table shows the volume of hydrogen gas produced over fifty seconds.

Time/s	Volume of gas/cm^3
0	0
10	20
20	40
30	55
40	65
50	72

The average rate at which gas is produced can be calculated as shown.

average rate between 10 and 20 seconds $= \dfrac{\text{change in volume of gas during time period}}{\text{length of time period}}$

$$= \frac{40-20}{20-10} = \frac{20}{10}$$

$$= 2\ \text{cm}^3/\text{s}$$

Calculate the average rate at which gas is produced between **20 seconds** and **30 seconds**.

_____ cm^3/s   **1**

(3)

Marks | KU | PS

14. Saliva contains an enzyme which breaks down starch.

(a) Name the type of chemical reaction taking place when starch breaks down.

_____

1

(b) A student carried out an experiment to break down starch.

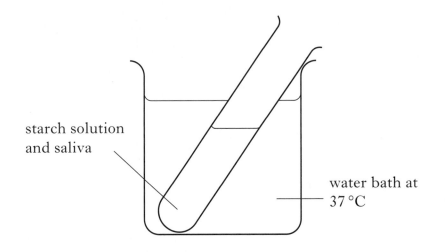

starch solution and saliva

water bath at 37 °C

He repeated the experiment using water at 100 °C.

What effect would this have on the activity of the enzyme?

_____

1

(c) The monosaccharide glucose is produced when starch is broken down.

Name another monosaccharide.

_____

1

(3)

Marks

15. Potassium hydroxide reacts with sulphuric acid to form potassium sulphate, which can be used as a fertiliser.

$$KOH(aq) \quad + \quad H_2SO_4(aq) \longrightarrow K_2SO_4(aq) \quad + \quad H_2O(\ell)$$

(a) Balance the above equation.   **1**

(b) Name the type of chemical reaction taking place.

_____   **1**

(c) Calculate the percentage, by mass, of potassium in potassium sulphate, K_2SO_4.

Show your working clearly.

_____ %   **2**

(d) Ammonium phosphate is also used as a fertiliser.

Write the **ionic** formula for ammonium phosphate.

1

(5)

[Turn over

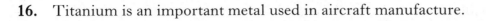

Marks | KU | PS

16. Titanium is an important metal used in aircraft manufacture.

(a) Titanium can be produced from titanium chloride as shown.

$$2Mg(s) + TiCl_4(\ell) \longrightarrow 2MgCl_2(s) + Ti(s)$$

Name the type of chemical reaction represented by the equation.

_____

1

(b) The magnesium chloride produced can be electrolysed as shown.

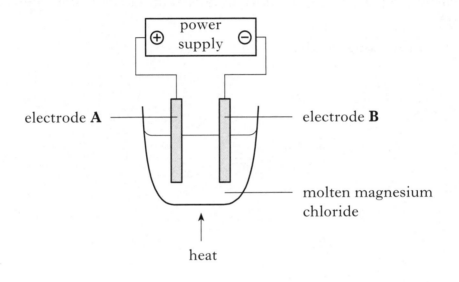

(i) At which electrode would magnesium be produced, **A** or **B**?

_____

1

(ii) Write the ion-electron equation for the formation of chlorine.
You may wish to use the data booklet to help you.

1

(3)

DO NOT WRITE IN THIS MARGIN

Marks | KU | PS

17. A solution of 0·1 mol/l hydrochloric acid has a pH of 1.

(*a*) (i) What colour would universal indicator turn when added to a solution of hydrochloric acid?

________________________________ **1**

(ii) Starting at pH 1, draw a line to show how the pH of this acid changes when diluted with water.

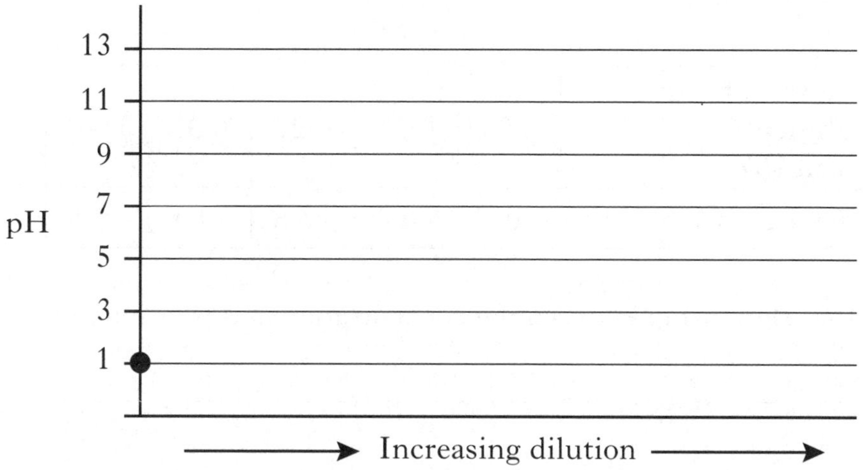

1

(*b*) Calculate the number of moles of hydrochloric acid in 50 cm^3 of 0·1 mol/l hydrochloric acid solution.

________________ mol **1**

(3)

[Turn over

Marks | KU | PS

18. A student investigated how the concentration of sodium chloride in water affected the freezing point.

(a) What type of bond is broken in sodium chloride when it dissolves in water?

_____

1

(b) The table shows information about the freezing point of different sodium chloride solutions.

Concentration of sodium chloride solution (mol/l)	0	0·09	0·18	0·27	0·37	0·46
Freezing point (°C)	0	−0·2	−0·5	−0·8	−1·1	−1·5

Describe the relationship between the concentration and freezing point.

_____

_____

1

(c) Predict the freezing point of a 0·55 mol/l sodium chloride solution.

_____°C   1

(3)

Marks | KU | PS

19. In Australia flow cells are used to store the energy from solar cells.

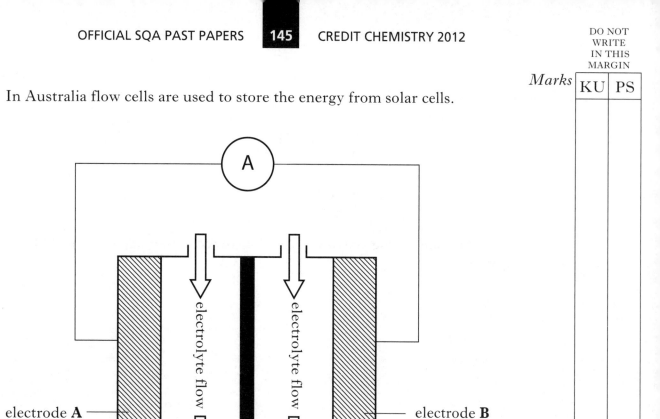

(a) The reaction taking place at electrode **A** when the cell is providing electricity is:

$$Zn \longrightarrow Zn^{2+} + 2e^-$$

Name the type of chemical reaction taking place at electrode **A**.

_____ 1

(b) **On the diagram**, clearly mark the path and direction of electron flow. 1

(c) Name the non-metal, that conducts electricity, which could be used as an electrode.

_____ 1

(3)

[Turn over

Marks | KU | PS

20. The monomer in superglue has the following structure.

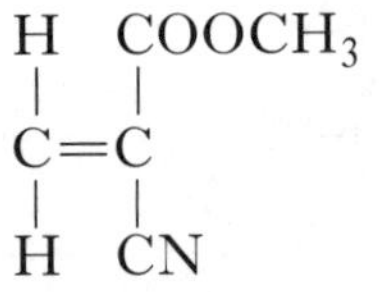

$$
\begin{array}{cc}
H & COOCH_3 \\
| & | \\
C & = C \\
| & | \\
H & CN
\end{array}
$$

(a) Draw a section of the polymer, showing **three** monomer units joined together.

1

(b) The polymer does **not** change shape on heating.

What term is used to describe this type of polymer?

_____

1

(c) Bromine reacts with the monomer to produce a saturated compound.

Draw the structural formula for this compound.

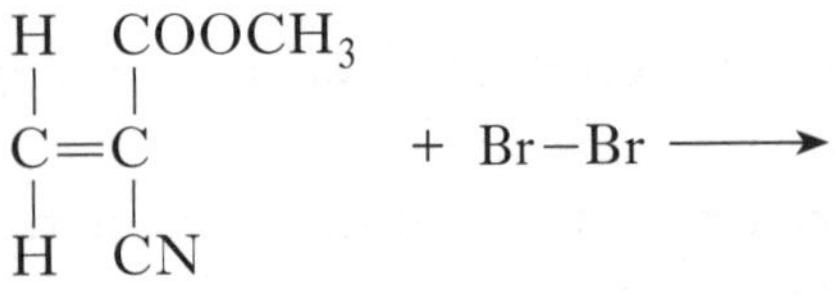

$$
\begin{array}{cc}
H & COOCH_3 \\
| & | \\
C & = C \\
| & | \\
H & CN
\end{array}
\quad + \ Br-Br \longrightarrow
$$

1

(3)

DO NOT WRITE IN THIS MARGIN

21. Aluminium is extracted from the ore bauxite.

(a) (Circle) the correct phrase to complete the sentence.

Aluminium is extracted from its ore $\left\{\begin{array}{l}\text{by heating with carbon}\\\text{by heating alone}\\\text{by electrolysis}\end{array}\right\}$.

1

(b) Aluminium can be mixed with other metals to make a magnet.

What term is used to describe a mixture of metals?

_____

1

(c) The composition of a 250 g magnet is shown.

Metal	aluminium	nickel	cobalt	copper	titanium	iron
% by mass	10	25	20	4	1	40

(i) Calculate the mass, in grams, of aluminium in the magnet.

Show your working clearly.

_____ g **1**

(ii) Using your answer to (c)(i), calculate the number of moles of aluminium in the magnet.

Show your working clearly.

_____ mol **1**

(4)

[END OF QUESTION PAPER]

ADDITIONAL SPACE FOR ANSWERS

ADDITIONAL GRAPH PAPER FOR QUESTION 12(*a*)

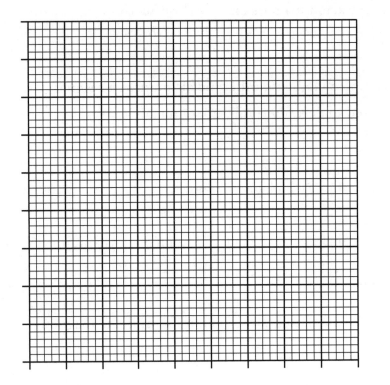

ADDITIONAL SPACE FOR ANSWERS

ADDITIONAL SPACE FOR ANSWERS

STANDARD GRADE | ANSWER SECTION

SQA STANDARD GRADE CREDIT CHEMISTRY 2008–2012

PART 1

1. (a) C   (b) B   (c) D

2. (a) A and F   (b) B

3. (a) C and D   (b) B and F

4. (a) D and F   (b) E   (c) C

5. (a) D   (b) A and C

6. D

7. (a) A and B   (b) E

8. C and E

9. B and E

PART 2

10. (a) Halogen

    (b) A covalent bond forms when two *positive* nuclei are held together by their common attaction for a shared pair of *electrons*.

    (c) As the size increases the amount of energy decreases / as the size decreases the energy increases **or** The energy required decreases as the size increases. / the energy required increases as the size decreases

11. (a) Hydroxide/OH^-

    (b) Increase/speed it up

    (c) Sacrificial

12. (a) (i)

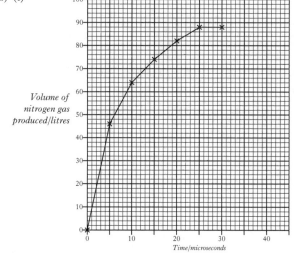

Volume of nitrogen gas produced/litres

Time/microseconds

      (ii)   13

    (b) $NaN_3 \rightarrow 1\frac{1}{2}N_2 + Na$

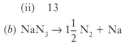

    or $2NaN_3 \rightarrow 3N_2 + 2Na$

    (c) Unreactive/will not react with sodium/does not burn/not flammable

13. (a) Conducts electricity/conductor

    (b) $2Cl^- \rightarrow Cl_2 + 2e$

    (c) Ions are free to move

14. (a) Carbon, hydrogen, oxygen

    (b) Sucrose

    (c) 0·0033

15. (a) (i)   1

      (ii)   Electron

    (b) (i)   proton 90

             neutron 144

      (ii)   84

16. (a) 68·3

    (b) Aluminium - electrolysis of molten compound Lead - using heat and carbon

    (c) Less reactive/lower in reactivity series

    (d) Reduction

17. (a) (i)   No reaction

      (ii)   Reaction occurred

    (b) (i) NO_3^-

      (ii)   Filtration/filtering or correct description

18. (a) Lightning

    (b) Bacteria **or** nitrogen-fixing bacteria **or** nitrifying bacteria

    (c) Ammonia/NH_3

    (d) As the temperature increases/*decreases* the solubility decreases/*increases* **or**

    The solubility increases/*decreases*, as the temperature decreases/*increases*.

19. (a)

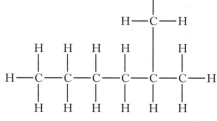

    (b) Isomers

    (c) 20–27

20. (a) 28 g

    (b) 3^+/three positive/Fe^{3+}

    (c) (i)   *Any one from:*
          • To provide/supply oxygen
          • To produce carbon dioxide/monoxide
          • To provide oxygen to react with carbon
          • For combustion to take place

      (ii)   Iron would be solid/harden/solidify Iron would not be able to flow/molten/liquid

CHEMISTRY CREDIT 2009

PART 1

1. (a) A   (b) F

2. (a) D   (b) F

3. (a) B   (b) D

4. (a) C   (b) A and C   (c) F

5. (a) B and E   (b) C   (c) B and F

6. (a) E   (b) C   (c) B and D

7. A and F

8. C and D

PART 2

9. (a) 10   10

       10   11

       10   12

    (b) Isotopes

    (c) 20 or ^{20}Ne

10. (a) Electrolysis

    (b) $(Al^{3+})_2(O^{2-})_3 Al_2^{3+}O_3^{2-}$

    (c) Ions can move/free ions
       Ionic lattice broken up

11. (a) Hydrogen or H or H_2
       Carbon or C

    (b) Test tube A – colourless liquid or water or H_2O
                 – condensation
       Test tube B – lime water did not turn milky or no change

12. (a) Covalent

    (b) Diagram must show 1 P and 3 H atoms
       3 pairs of shared electrons and 2 non-bonded electrons

13. (a) Neutralisation

    (b) (i) Gas given off/CO_2 given off

       (ii) Both scales correct
            Both labels correct including units
            Plots correct
            Joining points

    (c) Greater than 0.8
       No higher than 0.86

14. (a) (i) Ostwald

       (ii) Reaction is exothermic/gives out heat
            It is exothermic

       (iii) Water/H_2O

    (b) FM = 80

$$\% = \frac{28 \times 100}{80} = 35$$

15. (a) Y   X   W   Z

    (b) Hydrogen/H_2

    (c) Y = K/Na/Li/Ca/Mg
       Z = Hg/Ag/Au/Pt

16. (a) Hydrolysis

    (b) $C_6H_{12}O_6 \longrightarrow 2\,C_2H_5OH + 2\,CO_2$

    (c) distillation
       evaporation **and** condensation

    (d) FM = 46

$$\frac{230}{46} = 5$$

17. (a) Oxidation

    (b) Right to left (Y to X)
       On or very close to wire
       Not nearer to ion bridge

    (c) Add starch
       Blue-black

    (d) $Br_2 + 2e \longrightarrow 2Br^-$

18. (a)

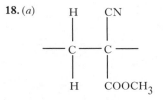

    (b) Carbon = 8 hydrogen = 17

    (c) Carbon monoxide or CO
       Hydrogen cyanide or HCN

19. (a) 216 – 221 inclusive

    (b) Homologous

    (c) 1 mol $\longrightarrow$ 10 mol

       128 $\longrightarrow$ 180

       6·4 $\longrightarrow$ $\dfrac{6 \cdot 4 \times 180}{128} = 9$

    (d) Correct structural formula for butane

CHEMISTRY CREDIT
2010

PART 1

1. (a) E   (b) A

2. (a) E   (b) A and C   (c) D and F

3. (a) A and D   (b) D

4. (a) E   (b) A and B

5. (a) B   (b) D

6. (a) B and F   (b) B   (c) E and F

7. (a) D   (b) C

8.   B and E

9.   A and F

PART 2

10. (a) Human made **or**
    Man made/unnatural/not natural

    (b) (i)

    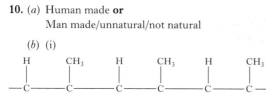

    (ii) Addition

    (c) Carbon monoxide/CO

11. (a) (i) Fermentation/anaerobic respiration
    (ii) Carbon dioxide/CO_2

    (b) The balloon does not inflate/inflates less/not as much

12. (a) Burns with a pop

    (b)

    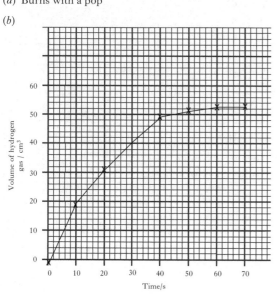

    (c) 41 +/-1

    (d) 53

(e)   1 mol   1 mol
      24·5   2
      4·9   0·4

or

n =   m/gfm
   =   4·9/24·5
   =   0·2

1 mol → 1 mol
0·2 mol → 0·2 mol

m =   n × gfm
    =   0·2 × 2
    =   0·4 g

13. (a) To allow the products to be identified/separated/ only one product/colour is collected at an electrode
    or
    To allow colours to be separated/identified

    (b) Green
    (c) Ni^{2+} CrO_4^{2-}

14. (a) Covalent

    (b) (i) $TiCl_4 + 4Na \rightarrow Ti + 4NaCl$
    (ii) Titanium is less reactive **or**
    Sodium is more reactive

15. (a) As the distance from the surface *increases* the concentration *increases.*/As the distance from the surface decreases the concentration decreases.
    or
    The concentration increases as the distance increases./ The concentration decreases as the distance decreases.

    (b) 80·5
    (c) (protons)   (electrons)   (both required)

16. (a) (i) 2+/two positive   **or**
    +2/positive 2
    (ii) gfm = 239
    207/239 × 100 = 86·6
    86·6% (87%)

    (b) (i) Iron/Fe
    (ii) Mercury   Hg
    Aluminium   Al
    Copper   Cu   (all 3 required)

17. (a) (NO_3^-)

    (b) Ferroxyl indicator turns blue

    (c) $Ag^+ + e^- \rightarrow Ag$
    $2Ag^+ + 2e^- \rightarrow 2Ag$

    (d)

18. (a) Ethoxypropane

    (b) (approximately) 36°C (32–40°C)

19. (a) 25 cm^3

(b) n = c × v
= 0·1 × 0·025
= 0·0025 mol HCI

$$1 : 1$$
$$0\cdot0025 : 0\cdot0025 \text{ mol NaOH}$$

c = n/v

$$= \frac{0\cdot0025}{0\cdot01}$$

= 0·25 mol/l NaOH

c = $\frac{n/m}{v}$ = zero

or

$$OH^- \times v \times c = H^+ \times v \times c$$
$$1 \times 10 \times \text{conc} = 1 \times 25 \times 0\cdot1$$
$$\text{conc} = \frac{1 \times 25 \times 0\cdot1}{1 \times 10}$$
$$= 0\cdot25 \text{ mol/l}$$

20. (a) (i) Propanol **or** propan-l-ol

(ii)
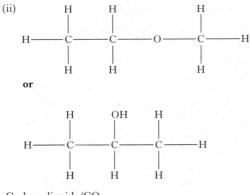

or

(b) Carbon dioxide/CO_2
Water/H_2O/hydrogen oxide

CHEMISTRY CREDIT 2011

PART 1

1. B

2. (a) E   (b) A

3. (a) A   (b) B and C   (c) C and D

4. (a) F   (b) A and F

5. (a) B and F   (b) C   (c) E

6. (a) D   (b) B

7. C and E

8. D and E

9. A and D

10. B and E

PART 2

11. (a) Distillation/fractional distillation
Evaporation then condensation

(b) (i) Isotopes

(ii) 8   10
8   8

12. (a) Fermentation
Anaerobic respiration

(b) (i) As the percentage increases...the density decreases
As the percentage decreases...the density increases

(ii) 20

13. (a)
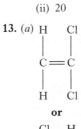

or

(b) Hydrogen chloride, HCI
Carbon monoxide

14. (a) a higher

(b)

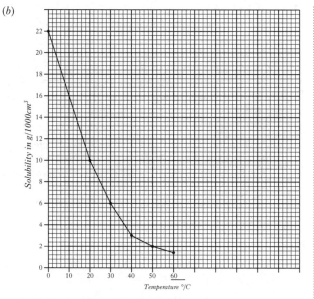

Both labels with units – ½ mark
Both scales – ½ mark
Plots correct – ½ mark
Plots joined – ½ mark

15. (a) $C_{12}H_{22}O_{11}$

(b) Maltose
Lactose

(c) (i) Biological catalyst

(ii) pH
Acidity or Alkalinity
Concentration of acid
Concentration of alkali

16. (a) Full or shortened structural formula of cyclopropane
eg

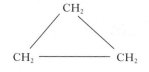

(b) (i) Lower temperature
Less heat/energy

(ii) Al_2O_3
If ion charges are shown all must be correct
$(Al^{3+})_2 (O^{2-})_3 / Al^{3+}_2 O^{2-}_3$

17. (a) hydrolysis

(b) 1mol   2mol
60   34
90          $\dfrac{90 \times 34}{60} =$ 51

18. (a) d.c. or direct current

(b) Chlorine gas
Bubbles of gas
Gas given off
Fizzing/effervescence
Green/yellow gas
Cl_2 (g)

(c) Two positive,   2+, Co^{2+}

19. (a) Arrows drawn from unreacted gases to hydrogen and
nitrogen box or catalyst box or between these two

(b) (i) Platinum, Pt

(ii) It is an exothermic reaction
The reaction produces heat

(c) Convert pollutant gases to harmless gases
Convert harmful gases to harmless gases

$CO \longrightarrow CO_2$ **or**
$NO_X \longrightarrow N_2$

20. (a) (i) $Pb(NO_3)_2(aq) + 2NaI(aq) \longrightarrow PbI_2(s) + 2NaNO_3(aq)$

(ii) filtration

(b)   Copper carbonate $CuCO_3$

(c) (i) Indicator/named acid/base indicator

(ii) Moles $n = c \times v$
$= 0{\cdot}1 \times 0{\cdot}02$ moles
$= 0{\cdot}002$ moles

(d) Apply mole ratio
$0.002 \longrightarrow 0.004$

21. (a) A $\longleftarrow$ B

(b) $Au^+(aq) + e^- \longrightarrow Au(s)$

(c) Ion bridge/salt bridge
Filter paper soaked in salt solution/electrolyte

22. (a) Same general formula and same/similar properties/same/
similar chemical properties

(b)

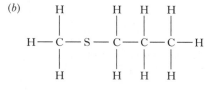

(c)   addition

CHEMISTRY CREDIT 2012

PART 1

1. (a) A and E   (b) D

2. (a) E   (b) C

3. B and C

4. (a) E   (b) D   (c) A and C

5. (a) F   (b) C and D   (c) D

6. (a) C   (b) B

7. (a) E   (b) D and F   (c) D

8. (a) A   (b) C and E

9. B and F

PART 2

10. (a) Stop air, oxygen and/or water, moisture, rain

    (b) (i) A / Zinc

       (ii) D / Silver

11. (a) 2

    (b) (i) Larger particle size, smaller surface area
         Large lumps

       (ii) 0.5g zinc

12. (a) Both labels with units   ½ mark
       Both scales   ½ mark
       Plots correct   ½ mark
       Plots joined   ½ mark

    (b) If graph drawn answer must be from graph
       If no graph drawn answer is 25.5 ±1

    (c) Speed up reaction, too slow at 200°C

13. (a) Any suitable diagram showing two hydrogen atoms with
       two electrons in the overlapped area

$$H \underset{X}{\overset{X}{\cdot}} H \qquad H \underset{\circ}{\overset{\circ}{\cdot}} H \qquad \left(\!H \underset{\circ}{\overset{\circ}{\cdot}} H\!\right)$$

    (b) (i) 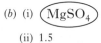 $MgSO_4$

       (ii) 1.5

14. (a) Hydrolysis

    (b) Less/no activity

    (c) Fructose

15. (a) $2KOH + H_2SO_4 \longrightarrow K_2SO_4 + 2H_2O$

    (b) neutralisation

    (c) FM = 174g
       $78/174 \times 100 = 44.8$
       44.8 or 45 on its own will also gain full marks

    (d) $(NH_4^+)_3PO_4^{3-}$

16. (a) displacement
      redox

    (b) (i) B/negative
       (ii) $2Cl^- \longrightarrow Cl_2 + 2e$
          $2Cl^- - 2e \longrightarrow Cl_2$

17. (a) (i) Red, pink, orange, yellow
       (ii) Line must be increasing and stop at 7 or below

    (b) $n = c \times v$
       $n = 0.1 \times 0.05$
       $n = 0.005$ moles

18. (a) Ionic
      Ionic lattice
      Ionic network

    (b) As concentration increases/decreases freezing point
       decreases/increases
       The freezing point decreases/increases as concentration
       increases/decreases
       As concentration increases freezing point gets colder

    (c) −1.8 to −2.0 inclusive

19. (a) oxidation

    (b) Left to right indicated on or near the wire

    (c) C, graphite, carbon

20. (a) Diagram must show three monomer units linked together

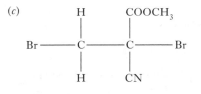

    (b) thermosetting
       thermoset
       thermal setting

    (c)

$$Br \!-\!\!-\!\!-\! \underset{\underset{H}{|}}{\overset{\overset{H}{|}}{C}} \!-\!\!-\!\!-\! \underset{\underset{CN}{|}}{\overset{\overset{COOCH_3}{|}}{C}} \!-\!\!-\!\!-\! Br$$

21. (a) by electrolysis

    (b) alloy

    (c) (i) 25g
       (ii) $(25/27) = \frac{1}{2}\, 0.926 / 0.93 \times \frac{1}{2}$
         0.926 / 0.93 or 0.9 on its own will also gain full marks

Hey! I've done it

BrightRED
PUBLISHING

© 2012 SQA/Bright Red Publishing Limited, All Rights Reserved
Published by Bright Red Publishing Ltd, 6 Stafford Street, Edinburgh, EH3 7AU
Tel: 0131 220 5804, Fax: 0131 220 6710, enquiries: sales@brightredpublishing.co.uk,
www.brightredpublishing.co.uk

Official SQA answers to 978-1-84948-239-4
2008-2012